CARBON MONOXIDE

Publication Number 1059
AMERICAN LECTURE SERIES®

A Monograph in
The BANNERSTONE DIVISION *of*
AMERICAN LECTURES IN ENVIRONMENTAL STUDIES

Edited by
CHARLES G. WILBER, Ph. D.
Department of Zoology-Entomology
Colorado State University
Fort Collins, Colorado

CARBON MONOXIDE

The Silent Killer

By

ROY J. SHEPHARD M.D., Ph.D.

Director of the School of Physical and Health Education
Professor of Applied Physiology
Department of Preventive Medicine and Biostatistics
Faculty of Medicine
University of Toronto
Toronto, Ontario, Canada

With a Foreword by

CHARLES G. WILBER, Ph.D.

Department of Zoology-Entomology
Colorado State University
Fort Collins, Colorado

CHARLES C THOMAS • PUBLISHER

Springfield • Illinois • U.S.A.

Published and Distributed Throughout the World by
CHARLES C THOMAS · PUBLISHER
2600 South First Street
Springfield, Illinois 62717, U.S.A.

ISBN 0-398-04850-9
Library of Congress Catalog Card Number: 83-635

With THOMAS BOOKS careful attention is given to all details of manufacturing and design. It is the Publisher's desire to present books that are satisfactory as to their physical qualities and artistic possibilities and appropriate for their particular use. THOMAS BOOKS will be true to those laws of quality that assure a good name and good will.

Library of Congress Cataloging in Publication Data

Shephard, Roy J.
Carbon monoxide, the silent killer.

(American lecture series; publication no. 1059)
"A monograph in the Bannerstone Division of American lectures in environmental studies"—P. ii.
Bibliography: p.
Includes index.
1. Carbon monoxide—Toxicology. 2. Carbon monoxide—Physiological effect. I. Title. II. Series. [DNLM: 1. Carbon monoxide poisoning. QV 662 S548c]
RA1247.C17S48 1983 615.9'1 83-635
ISBN 0-398-04850-9

Printed in the United States of America
PS-R-1

Heav'n's youngest teemed star
Hath fixt her polisht car

Milton

Southern United States people will be uncomfortable if heating fuel is cut off, but in Canada many people would die.

J. Tuzo Wilson

INTRODUCTION

My acquaintance with carbon monoxide is long-standing. As a small boy, I recall the fascination of "fetching the taper," a burning brand on a long pole, used by our teacher to ignite the gas lamps for the late afternoon classes at our school. Father had "modern" ideas, condemning our family to the unproved methodology of electric cooking, heating, and laundering. But beyond our home, gas reigned supreme. On the annual pilgrimage to the North London Exhibition, we would stroll slowly under the flickering jets that lit the vast Victorian caverns of Alexandra Palace, the air heavy with the odour of coal gas. Fortunately, the products pumped from the gasworks of North London had a characteristic, pungent, and quite unpleasant odour. The carbon monoxide content of the Palace and other public buildings probably exceeded modern standards, but it is unlikely that anyone with a normal sense of smell succumbed to clinical poisoning.

It might seem that the gas era has passed. My childhood memories include the excitement of watching shops, public buildings, and streets progressively converted from gas to electric lighting. Nevertheless, new sources of carbon monoxide — particularly automobile exhaust — have actually increased the atmospheric burden of carbon monoxide in urban areas. The emission controls fitted to cars of the 1970s temporarily checked the rising curves of air management agencies, but it seems likely that the benefit from such measures will soon be outweighed by the burgeoning vehicle ownership of a rising world population. Unlike the coal gas of an earlier generation, the urban pall of carbon monoxide lacks a distinctive warning odour — it has become a silent killer. Moreover, we can no longer escape by taking a walk in the Palace gardens; the pall is thickly spread over several hundred square kilometres of a modern megalopolis.

The energy crisis of the 1980s threatens a further increase in exposure to carbon monoxide. The improved sealing of our homes to hot and cold weather exposes the occupants to the full perils of emissions from cigarettes, heating and cooking appliances, and paint strippers. Out of doors, also, demands for economy of gasoline consumption are inducing the delay and in some instances the relaxation of controls upon automotive emissions.

Lastly, continuing smokers have finally realized it is not too clever an idea to pump carcinogenic tars into their lungs forty or more times per day. Rather than relinquish their oral fixation, many addicts have turned to low tar, low nicotine cigarettes. Again, the silent killer is overlooked. A suburban housewife can wax eloquent about the evils of air pollution while cheerfully inhaling 4% carbon monoxide from her cigarette; indeed, by choosing the trendy low tar, low nicotine variety, she may well encounter much more carbon monoxide than a less enlightened smoker, as she seeks her daily "fix."

So carbon monoxide is a persistent and indeed a growing contaminant of the air breathed by the average city dweller, and it is timely to make a critical review of the health consequences of exposure to this pollutant. The main volume of research upon carbon monoxide has been undertaken over the last fifteen years, and this activity is here represented by a fairly comprehensive bibliography of some 1200 entries. However, as in other areas of pathophysiology, a number of the more fascinating anecdotes are provided by the investigators of an earlier era. Modern committees on human experimentation would look askance at experiments where the standard end point was the subject's loss of consciousness, but thanks to the intrepid work of men like J. S. Haldane, we have detailed and accurate descriptions of events associated with massive carbon monoxide exposure. I have thus thought it appropriate to set the more recent findings in their historical context.

As is inevitable when writing a monograph of this sort, I have drawn heavily upon information gained from my colleagues, both distinguished and more junior. The matter of urban and industrial standards has been the subject of deep discussion with officials of Health and Welfare Canada (particularly Ms. Betty Meek) and of the provincial government (particularly Dr. Max Fitch and Mr. Lou Shenfield). Domestic contamination was reviewed with a committee of United States environmentalists headed by Dr. Arthur Dubois. Much of the detailed surveying of the literature and the research of our laboratory on such matters as urban contamination, driving hazards, and passive smoking was undertaken by a group of enthusiastic and talented graduate students, including Dr. Geoffrey Wright, Gaetan Godin, Peter Pimm, and Dr. Andris Rode, with strong technical assistance from Brigitte Berhens, John Sebben, and Elenor Ponsford. Lastly, the deciphering of my oft-amended notes was left to the competent hands of Jocelyn Lee.

My older daughter, Sarah, shared in the writing of an earlier book (*Endurance Fitness*, 1969) through an essay on the theme "Cats are Lovely." It was a particular thrill this year to enjoy a rigorous criticism of the section on the chemistry of carbon monoxide from a young lady who is now establishing her reputation as an organic chemist under the tutelage of Professor Arigoni in Zurich. My younger daughter, Rachel, is showing a strong bent for environmental studies, and if a further edition of this book is justified, it may well become a family project. My wife, Muriel, is perhaps less familiar with atmospheric chemistry but has also brought her special gifts to the author—not least the tranquil domestic environment and the frequent liquid refreshment needed in tackling a dry subject.

A heavy debt of gratitude is owed to all of these individuals and to others who in various ways have contributed to the realisation of this volume. The book, in itself, may seem a poor repayment, but if a consequence of its publication is the breathing of cleaner air, I shall be happy to have forwarded one urgent concern of these donors.

ROY J. SHEPHARD

CONTENTS

CARBON MONOXIDE

Chapter 1

CARBON MONOXIDE YESTERDAY AND TODAY

HISTORICAL BACKGROUND

Classical concepts of the breathing processes were so bizarre (Shephard, 1977) that it is difficult to be certain how far the effects of carbon monoxide poisoning were understood. Lewin (1920) noted some writing by Aristoteles in the third century BC, and by the time of Cicero (106-43 BC), charcoal fumes were an accepted agent for both suicide and the execution of criminals (Lewin, 1920; Cooper, 1966).

The use of coal for domestic heating became increasingly popular in the fifteenth and sixteenth centuries (Air Pollution Control Association, 1970), creating problems for both the homeowner and the miner. Deacon (1616) noted the intent of chimneys, "that thereby, the smoke . . . might be dispersed abroad in the aire for fear of hurting the bodies of men." Paracelsus (1493-1541), lecturer in medicine at the University of Basel and an alchemist deeply committed to the search for the "philosopher's stone" that would turn all matter to gold, was the first person to write a treatise on diseases afflicting miners. His understanding of mine gases was hardly helped by a tripartite concept of matter (mercurial or volatile, sulphureous or combustible, and saltlike or residual). Van Helmont (1577-1644) also discussed risks to miners, attempting to distinguish differences in the temperament of the various "gases" to which they might be exposed; for example, gas sylvestre (carbon dioxide) had a wild, chaotic quality that extinguished a naked flame. Once, when trying to write in a small room with no ventilation, he was personally attacked by the fumes of "wood gas" from a pot of charcoal and barely escaped from his study before falling down "half dead." Ramazzini (1633-1714), a physician and professor at the University of Modena, wrote a more comprehensive manual entitled *De Morbis Artificum (Diseases of Workers,* Wright, 1964). He recognized that many of the materials encountered by the workers emitted "noxious vapours" and would have welcomed the provision of more effective gas masks. In the case of the miner, "when those vapours have gained entrance to the dwelling house of life and are mixed with the blood, they pervert and pollute the natural composition of the nervous fluid, and the result is palsy, torpor, and the maladies above mentioned . . . ." Ramazzini also commented on the hazard of burning coals as encountered by confectioners preparing almonds and pistachio nuts; a combination of heat and fumes from the burning coals and molten sugar caused serious complaints, including headaches, pains in the eyes, and severe dyspnoea. The burning coals had a "potent pernicious quality, which unless you give it a free outlet to escape by can kill a man almost on the instant."

Towards the end of the nineteenth century, a series of mine disasters in England was investigated by the celebrated respiratory physiologist J. S. Haldane (Haldane and Priestley, 1935). He soon established that 95% of the miners died not of burns or mechanical injuries but rather of carbon monoxide poisoning. Analysis of air samples from contaminated shafts did much to clarify understanding of the various "damps" (Middle German *dampf*= vapour) described by earlier generations of miners. *Black damp* was formed during the spontaneous oxidation of coal or timber. Its main constituents were carbon dioxide and nitrogen; it thus provoked an increase of breathing if present in moderate concentrations but depressed respiration when in undiluted form (12-15% CO_2). The carbon monoxide content became significant if there was any increase in temperature of the coal. It was thus evident that carbon monoxide sensing equipment not only would protect the miners against poisoning but also would provide an early warning of fires (Graham, 1920-21; Haldane and Makgill, 1933). *White damp* was formed during the spontaneous oxidation of wet coal (both in mines and in the holds of coal-carrying ships); a mixture of carbon dioxide and carbon monoxide was given a whitish appearance by the presence of condensed moisture. *Fire damp* consisted largely of methane. The main risk from this contaminant was an explosion, but since the gas was irrespirable, it also increased the respiratory danger arising from carbon monoxide poisoning. The main source of carbon monoxide was *after damp*, the deadly gas remaining after a fire or explosion. A sample of pure after damp, collected from the Altofts colliery, contained 11.9% carbon dioxide, 8.6% carbon monoxide, 2.9% hydrogen, 3.1% methane, and 73.5% nitrogen (Wheeler, 1910). Even if the oxygen supply remained adequate, men who had been exposed to after damp became reckless, pushing forward into the contaminated workings until halted by a faintness of "failure of the legs," which left little possibility of escape. In many fatal incidents, the carbon monoxide concentration was no more than 2%, yet the final blood carboxyhaemoglobin readings of the victims were as high as 80%. Haldane (1895b) suggested that rescue workers should carry a mouse or a canary with them; because of a large relative respired gas volume and a high rate of metabolism, such a creature would show distress before the rescuer was dangerously affected. He noted, "While recently collecting samples of the gases from the

Staffordshire collieries, I have used mice with this end in view, and found that their symptoms gave timely warning of the presence of carbon monoxide in poisonous proportions."

Haldane went on to develop and to test various types of breathing equipment for use in contaminated atmospheres; these generally included a cylinder supply of pure oxygen and, in one United States device, a catalyst that oxidized carbon monoxide to the much less dangerous carbon dioxide. A further important contribution was to recommend the modification of mine ventilating shafts to allow a reversal of airflow in the event of a fire.

Back in his laboratory, Haldane exposed himself and his friends to what would now be regarded as dangerously high concentrations of carbon monoxide (leading to blood carboxyhaemoglobin levels of as much as 56%). Above 50% saturation "there was no real discomfort at rest, but . . . vision and hearing [were] impaired and intelligence [was] probably greatly impaired." He noted that recovery was speeded by the breathing of oxygen, preferably in combination with some carbon dioxide, although he admitted that (at least in the case of miners who were resuscitated) the "nervous system generally is likely to remain very seriously affected for days, weeks or months . . . Mental powers and memory may be much impaired."

Coal was distilled into gas and coke by a scientifically minded clergyman, the Reverend John Clayton, in 1688, but the use of gas as an illuminant was first proposed by William Murdock in 1792. A Birmingham factory was lighted by this means in 1798, and the first public gasworks was constructed at Salford in 1805. In the face of much ridicule and some apprehension, London began to install gas street lighting in 1808. As the nineteenth century progressed, an ever more widespread use of manufactured gas (see Table I) increased the potential for both suicide and murder while causing many industrial and domestic accidents. Typical histories included fracturing gas lighting brackets by the weight of a heavy coat, extinguishing a metered gas fire while the tap remained open, and blocking the vents to a space or water heater (often in an attempt to conserve heat). Haldane (1899) chaired a committee charged with the investigation of this problem. He commented on factors predisposing to accidents in the home, including small room size (with a rapid accumulation of toxic gas concentrations), solitary occupancy (reducing the chances of detecting a leak), and the frequent use of coin gas meters. The ventilation of a room was often a critical determinant of whether a fatal accumulation of gas occurred. Since carbon monoxide was a little lighter than air, the toxic material usually layered near the ceiling of a room, sometimes imposing a hazard upon the occupants of the floor above. However, if the room was cooling, convection currents might carry the carbon monoxide back towards the floor.

The committee found that fatalities were more frequent in the United States than in Britain, and this was traced to a difference in gas technology.

Table I
COMMON SOURCES OF CARBON MONOXIDE*

Source	Carbon monoxide concentration (%)
Water gas	43.6
Producer gas	33.5
Blast furnace gas	26.2
Coal gas	7.4
Coke gas	5.1
Vehicle exhaust	8-12
Cigarette smoke	2-4

*Based in part on Wade, 1964.

In Britain, illuminating gas was usually produced by the distillation of bituminous coal; the main constituents were methane (29%) and hydrogen (52%), while the carbon monoxide concentration was only 7 to 8% (Wade, 1964). In contrast, some United States cities were generating much more toxic forms of gas (Drinker, 1938). Producer gas (25 to 34% CO, with much of the residue as nitrogen) was formed by passing air through a bed of incandescent coke, while carburetted water gas (40 to 44% CO) was manufactured by blowing steam through coke heated to 1100° C and adding some unsaturated hydrocarbons. Coal gas ($21 MJ \cdot m^{-3}$) proved a better source of heat than producer gas ($4.4\ MJ \cdot m^{-3}$) or water gas (up to $12.3\ MJ \cdot m^{-3}$). Because of the United States accident experience, it was decided to prohibit the distribution of gas containing more than 30 per cent carbon monoxide in Britain (Haldane and Priestley, 1935). Odiferous substances were also added to the gas to increase the likelihood of early discovery of a leak. Unfortunately, the most vulnerable group of gas users, absentminded senior citizens, also has the poorest sense of smell. The use of coal gas continued in New York City until 1956 (Eisenbud and Ehrlich, 1972) and in England until the North Sea discoveries of natural gas in the 1960s. Malleson (1973) noted a dramatic drop in the suicide rate for Britain commencing about 1963; although credit for this trend was claimed by both psychiatrists and a telephone counselling service called the "Samaritans," Malleson suggested that accolades should in truth be directed to the Gas Board, whose natural gas conversion programme had removed one of the most common suicide weapons.

During World War II, drivers were startled to see trucks with large balloons on their roofs and buses towing small furnaces. A shortage of gasoline had forced the conversion of many commercial vehicles to the use of producer gas. Although there were not many instances of acute poisoning, workers began to complain that the fumes were causing chronic carbon monoxide poisoning (Grut, 1949; Petri, 1953; Zorn and Kruger, 1960; Pfrender, 1962; Breysse, 1961; Bell, 1961.).

Subsequent to World War II, occasional accidents have continued from wood, charcoal, and paraffin stoves, particularly in temporary homes on Canada's northern frontiers. Interest in industrial hazards has also continued, but the main focus of concern has shifted to the lower concentrations of carbon monoxide encountered on crowded highways due to an accumulation of vehicle exhaust fumes and in poorly ventilated rooms due to the accumulation of cigarette smoke. Both of these problems are still with us and indeed are an urgent concern of the environmentalist. The pressures of urban growth are causing a progressive increase in the density of vehicular traffic, while the storage of fossil fuels is creating some agitation to disregard automotive design improvements that reduce carbon monoxide emissions at the price of greater gasoline consumption. Likewise, the domestic architect is under pressure to conserve both heat and refrigeration by reducing indoor ventilation; the resultant accumulation of carbon monoxide —whether from smoking, industrial processes, or malfunction of central heating—is unfortunately a growing hazard of the indoor environment (Yocum et al., 1971).

CHEMICAL PROPERTIES

In 1772, the English chemist Joseph Priestley (1733-1804) described a combustible gas that burned with a bright blue flame quite different from the relatively clear flame of "common inflammable air" (hydrogen). However, carbon monoxide continued to be confused with hydrogen until 1800, when Cruickshank demonstrated that it was an oxide of carbon that could be converted to carbon dioxide by exploding it with oxygen.

The pure gas is colourless, odourless, and tasteless; since death can result from exposure to a concentration of 0.1% for as little as one hour or 0.05% for eight hours, it is an extremely dangerous compound. The density relative to hydrogen is fourteen, a little less than that of air (see Table II).

Table II
PHYSICAL PROPERTIES OF CARBON MONOXIDE*

Molecular weight	28.01
Density at 0°C, 1 atmosphere	$1.25\ g \cdot l^{-1}$
Specific gravity relative to air	0.967
Solubility in water at 0°C	$3.54\ ml \cdot dl^{-1}$
Explosive limits in air	12.5-74.2%
Fundamental vibration transition	4.67 μm

*Based on data collected by the National Academy of Sciences, 1977.

Other important properties are (1) a shorter interatomic distance than would be anticipated for a single bond (~1.13 Å), (2) a high heat of formation from the constituent atoms (bond strength 2.07 $MJ \cdot mol^{-1}$), and (3) a low dipole moment (implying that on average, the electrical charge is distributed rather uniformly between the carbon and the oxygen atoms). These observations led Pauling (1960) to suggest that in normal circumstances carbon monoxide existed as a hybrid containing roughly equal proportions of three structures:

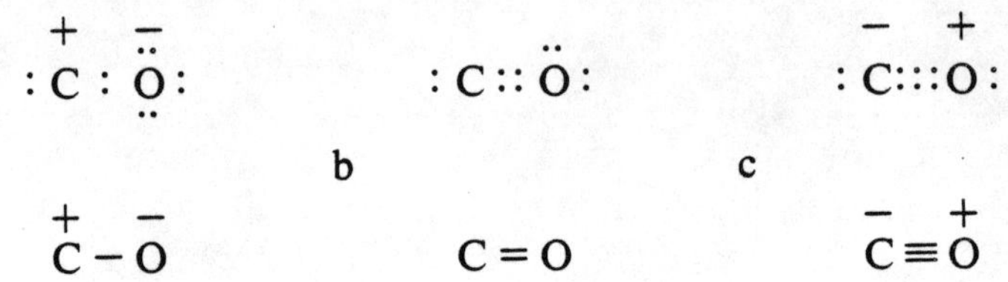

Modern atomic theory shows the carbon atom to have the first shell filled by two electrons in S orbitals, while the second shell contains two electrons in S orbitals and two in P orbitals.

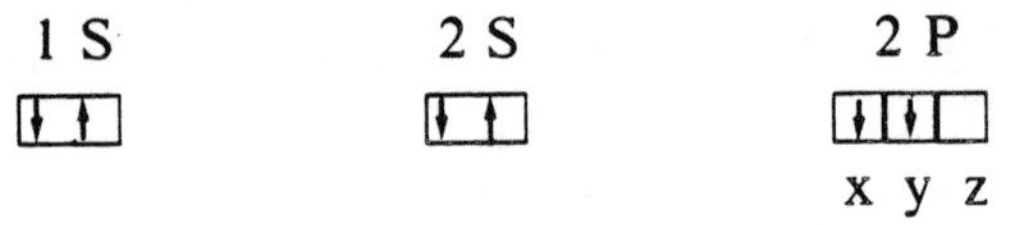

Likewise, for oxygen, the first shell contains two electrons in S orbitals and the second shell two electrons following S orbitals and four in P orbitals.

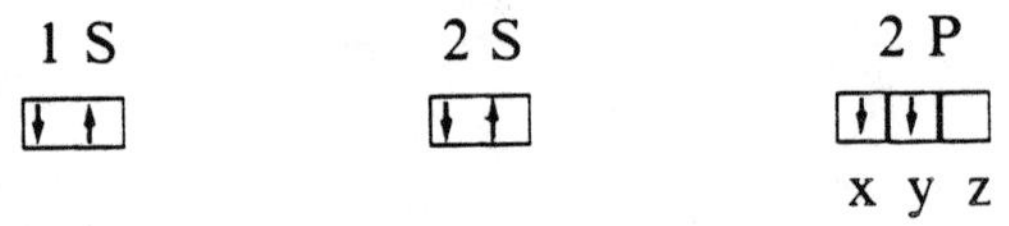

The conventional Lewis diagrams (as in a, b and c above) are now regarded as a somewhat inadequate representation of events. The initial bond is a hybrid *sigma bond* involving electrons from the S and P orbitals; at this stage, the greater proton mass of the oxygen holds the shared electrons closer to the oxygen nucleus, giving this a small negative charge ($\delta-$) relative to carbon ($\delta+$). The remaining two bonds are *pi bonds* formed by displacement and sharing of P orbitals; since the oxygen atom plays a dominant role in donation, it develops a small positive charge ($\delta+$) relative to carbon ($\delta-$).

Like other heteratomic gases, carbon monoxide has a strong and characteristic infrared spectrum. Since there are only two constituent atoms, there is but one possible mode of vibration, along the axis linking the carbon and oxygen atoms. Excitation is maximal at the resonant frequency for this bond, corresponding to a wavelength of 4.6 μm.

At normal room temperatures, carbon monoxide is stable and inert, but as the temperature is raised it becomes a good reducing agent and indeed is used to reduce metallic oxides in various smelting processes. Stoichiometric mixtures of carbon monoxide and oxygen explode in the presence of traces of moisture, methane, or hydrogen sulphide. Classical chemistry envisaged reaction of the type

$$CO + H_2O \longrightarrow CO_2 + H_2$$
$$2H_2 + O_2 \longrightarrow 2H_2O$$

More recent work has suggested that free radicals (hydroxyl and hydroperoxyl) provide low energy paths leading to the explosive reaction. The normal explosive limits in air are from about 12 to 75% CO. However, a mixture of carbon monoxide with air or oxygen cannot readily be exploded by an electric spark if the atmosphere is first dried thoroughly by an agent such as phosphorus pentoxide.

In the presence of a heated catalyst such as palladium, iron, or nickel, the disproportionate reaction occurs.

$$2\,CO \rightleftharpoons CO_2 + C + 162\ kJ \cdot mol^{-1}$$

The equilibrium proportion of carbon dioxide drops from 90% at 550° C to less than 1% at the 1100° C temperature used in gas manufacture.

Normally, almost all of the carbon monoxide is in its ground state. Even if heated to 1000° C, less than 1% of molecules have a P orbital electron displaced to a higher energy level. However, the energy of short-wave ultraviolet radiation (xenon, 129.5 nm) is capable of inducing this displacement. The higher energy forms react to yield various suboxides.

$$C + CO \longrightarrow C_2O$$
$$C_2O + CO \longrightarrow C_3O_2$$

A relatively narrow waveband of irradiation is effective in producing the reaction. Absorption of radiation at wavelengths of less than 111 nm dissociates the carbon monoxide further into its constituent atoms.

Compounds known as carbonyls are formed with some metals such as nickel—$Ni(CO)_4$—and iron—for instance, $Fe(CO)_4$. From the viewpoint of the physiologist, the attachment of carbon monoxide to the iron atoms of haemoglobin, myoglobin, and certain of the cytochrome pigments is of great practical importance.

When reacting with ferrous iron, the carbon atom of carbon monoxide forms a sigma bond, involving also a D-orbital electron from the third shell of Fe^{2+}. In the body, the Fe^{2+} also has four nitrogen linkages that contribute a substantial electron density; the effective valency is less than $^{2+}$, enabling more readily reversible reactions with both oxygen and carbon monoxide. Exposure to ultraviolet light ruptures the normal firm bonding of carbon monoxide with body pigments, at least in part because the dissociation product is converted to the suboxides.

Carbon monoxide has a relatively low water solubility (about 20% less than that of oxygen), and for this reason negligible quantities dissolve in physical solution in the blood at normally encountered partial pressures of the air contaminant.

NATURAL SOURCES

There has been vigorous debate concerning the relative importance of man-made and natural pollution of the atmosphere with carbon monoxide. One paper, popular with those who would decrease the regulation of human emissions, suggested that the natural formation of carbon monoxide from methane exceeded the anthropogenic output (Coburn, 1979; Jaffe, 1973; Levy, 1972; Maugh, 1972; McConnell et al., 1971) by a factor of ten (McCormick and Xintaras, 1962). More recent estimates, based upon the distribution of selected carbon monoxide isotopes in the atmosphere (Crutzen and Fishman, 1977; Seiler, 1978; Stevens et al., 1972) suggest that most of the atmospheric carbon monoxide is man-made. Moreover, human factors have clearly brought about the substantial gradient of carbon monoxide concentrations between rural areas (partial pressure 0.003 to 0.08 Pa,* depending upon the movement of local and continental air masses; Robbins et al., 1968; Junge, 1963) and major conurbations (background pressures 0.5 to 1.0 Pa, rising to 4 to 10 Pa on polluted streets).

The global average of CO pressures in the earth's atmosphere (the troposphere) is about 0.014 Pa in the northern hemisphere and 0.006 Pa in the southern atmosphere (Robinson and Robbins, 1970). There is some evidence that the burden is increasing, with important implications for the photochemistry of the stratosphere, including the stability of the ozone layer that protects man from excessive ultraviolet irradiation. Partial pressures drop from 0.008 Pa at an altitude of 4 km to 0.004 Pa at 15 km (Goldman et al., 1973). Carbon monoxide is produced by the spontaneous oxidation of organic matter and in natural conflagrations such as forest fires. It is a constituent of volcanic, marsh, and natural gases and is formed during electrical storms (Jaffe, 1968). Small quantities of CO are also liberated during the germination and growth of seedlings (Siegel et al., 1962) and the decomposition of green vegetation (Wilks, 1959). Larger amounts of carbon monoxide, up to 5% of that generated by combustion of fossil fuels, are produced by certain jellyfish (Barham and Wilton, 1964) and brown algae of the seashore (Chapman and Tocher, 1966).

The annual human output of carbon monoxide was estimated at 200 million tons in 1968 (Borg and Robinson, 1968), rising to 270 million tons in 1972

*In accordance with the recommendations of the International Committee on the Standardization of Units (Ellis, 1973), partial pressures are reported in Pascals (Pa). For most purposes, 10 parts per million (ppm) = 1 Pa = 11.5 mg · m^{-3}. However, at high altitudes a given partial pressure is equivalent to a higher volumetric (ppm) concentration. In the Canadian winter, the drop of environmental temperature also can reduce the volumetric reading due to an increase in density of the ambient air.

(Maugh, 1972). The latter is about 50% of the content of the troposphere. Nevertheless, background levels of CO remain relatively stable. The main mechanism for disposing of carbon monoxide seems to be the anaerobic soil bacteria; if hydrogen is available they reduce the CO to methane, and in the absence of hydrogen they oxidize CO to CO_2 (Inman et al., 1971). It is also possible that the CO reacts with hydroxyl radicals formed by the action of ozone upon water vapour in the upper atmosphere (Hampson and Garvin, 1975; Weinstock, 1969). The average residence time of a carbon monoxide molecule in the troposphere may thus be no longer than 0.2 years (Weinstock et al., 1972, 1974).

ANTHROPOGENIC SOURCES

The principal sources of ambient carbon monoxide in the United States are (in order of importance) transportation, forest and agricultural fires, industrial processes, the burning of solid waste, and the combustion of fuel in homes and factories (see Table III). The United States accounts for about a third of the global CO output (World Health Organization [WHO], 1977).

Transportation

Although gasoline-powered motor vehicles are by far the largest source of CO emissions in most industrialized nations (Jaffe, 1970), problems of CO accumulation were first described in connection with coal-fired steam locomotives. Haldane and Priestley (1935) noted that coal-burning engines were used on the Metropolitan section of the London underground railway. A special smoke-storing locomotive had been developed, but in the heyday of this system the air on the closed sections of the route contained up to 0.06% CO

Table III
PRINCIPAL AMBIENT SOURCES OF CARBON MONOXIDE IN THE UNITED STATES*

Source of carbon monoxide	Output (10^6 tons per year) 1940	1969†	1975
Transportation	35	112	73
Forest and agricultural fires	28	18	3
Industrial processes	14	12	15
Domestic and industrial fuel consumption	6	2	1
Solid waste disposal	2	8	4
Total	85	151	96

*Based on data of Babcock and Nagel, 1972; Council on Environmental Quality, 1971; and U.S. Environmental Protection Agency, 1975, 1978.

†1969 was apparently the peak year for automotive emissions in the United States.

(60 Pa, Haldane, 1897). The official enquirers were surprisingly unperturbed by these observations, concluding that the drivers of the trains were not exposed to the worst fumes for long enough periods to cause medical problems, while workmen repairing the tunnels usually restricted their operations to the night shift, when much of the carbon monoxide had dispersed. It was recommended that the route be electrified, but perhaps because of the gentle wording of the report the last steam locomotives were not taken from the Metropolitan route until the 1950s. More obvious difficulties had been encountered by drivers operating steam locomotives through the long tunnels of the Apennines. On these routes, a combination of carbon monoxide accumulation and low oxygen pressures had sometimes led to inefficient combustion, with dimming of lights in the passenger carriages and occasional loss of consciousness by the engine crew.

Currently, cars account for more than a half of the global output of carbon monoxide (Bowne et al., 1971). United States statistics for 1975 (Environmental Protection Agency, 1975, 1979) show an output of 63×10^6 tons derived from gasoline-powered vehicles operated on the highway and a further 7×10^6 tons emitted by motor vehicles when off the highway. In contrast, other forms of transport are relatively minor sources of pollution (shipping 1.1×10^6 tons, aircraft 0.9×10^6 tons, diesel vehicles, 0.6×10^6 tons, and railroads 0.3×10^6 tons).

In the absence of emission controls, a car produces about 0.37 kg of carbon monoxide for each litre of gasoline that is consumed (an oxidation efficiency of ~ 92% and an output of ~ 50 g per km of road travelled by a full-sized North American car). The exhaust from a carburetted engine of this type contains some 4% by volume of carbon monoxide (Chipman and Massey, 1960; Elliott et al., 1955; Hurn, 1962; Larson et al., 1955; Twiss et al., 1955), in contrast with the 0.1% CO emitted by various classes of diesel vehicles (Elliott et al., 1955). This reflects the greater air to fuel ratio for diesel engine operation. On the urban routes of the United States, more than 80% of vehicle-kilometres are attributable to gasoline-powered cars and trucks (Babcock and Nagda, 1972; Bowne et al., 1971; WHO, 1979).

While the volume of exhaust fumes depends upon the engine speed, the CO concentration depends largely upon the air to fuel ratio (Air Pollution Control Association, 1970). Engine standards are specified for a steady speed of 32 $km \cdot h^{-1}$, and contaminant emissions can rise by a factor of up to four in the start-and-stop driving of a crowded city street; on the other hand, when cruising on a freeway at 70 to 100 $km \cdot h^{-1}$ emissions may drop to one-half of the standard 1975 average of 34 $g \cdot km^{-1}$ (Rose and Smith, 1962; Rose et al., 1965; National Academy of Sciences, 1977).

Concern about urban air pollution led to substantial modifications of engine design in the early 1970s. A reduction of as much as 70% in the emissions from a given vehicle was achieved by a substantial increase in the air to fuel ratio (to approximately 14.5 to 1.0, Air Pollution Control Association, 1970; Gates and Goldmuntz, 1976). Other expedients included preheating of the air intake, increase of idling speed, electronic control of fuel injection and ignition timing, arrangements for the recycling of exhaust, and (most recently) addition of catalytic converters for use with unleaded gasoline. This latest tactic makes a substantial contribution to control of CO emissions, but unfortunately the unleaded fuel is expensive and has the serious disadvantage of generating a potentially harmful sulphuric acid aerosol during combustion.

The intent of the United States Environmental Protection Agency (1979) is to reduce average emissions of carbon monoxide to ~ 13 g/km by 1995, as older and less well controlled vehicles are eliminated. In practice, a number of factors have conspired to reduce the favourable impact of the required engineering changes upon the quality of the urban environment (Goldsmith and Landaw, 1968). The number of vehicles entering large cities has continued to increase by as much as 4 to 5% per annum. Performance statistics refer to new or regularly maintained engines, and the motoring public has been reluctant to pay for necessary 15,000 km operating checks (Chass et al., 1972). One survey (admittedly before the widespread use of electronic timing and fuel injection systems) found that 70% of vehicles entering a large North American city needed maintenance work on the carburetor system (Winthrop, 1972). Combustion of the gasoline becomes less complete and emission rates rise with high altitude (Larsen and Konopinski, 1961) and cold weather conditions (Hurn, 1962). In the absence of a catalytic converter, the carbon monoxide production is three times as great at −7° C as at 27° C (National Research Council, 1977). Particularly during short city trips, a substantial fraction of the total exhaust emission occurs before the engine has reached its optimum operating temperature (Hass and Brubacher, 1962).

Whether the 1995 standards will be realized remains problematic (see Table IV). Air pollution control devices are costly in terms of power output and fuel efficiency. The energy crisis has thus led to calls for relaxation of emission standards. At the same time, legislation requiring lighter and more fuel-efficient cars, the rapidly rising price of gasoline, and the introduction of alternative fuels may all reduce the amount of gasoline that is being consumed per annum.

The local street-level concentrations of carbon monoxide are often much higher than the general urban figure reported by control agencies. Street-level readings are markedly influenced by factors such as wind speed and direction, atmospheric stability, the local traffic density, and the height of nearby buildings. Figures for open areas are typically 1 to 5 Pa (Bové and Siebenberg, 1970; Brice and Roesler, 1966; Georgii et al., 1967; Godin et al., 1972; Ontario Ministry of the Environment, 1972; Ramsey, 1966; Wilkins,

Table IV
AUTOMOBILE EMISSION CONTROL SCHEDULES FOR UNITED STATES*

Schedule	Carbon monoxide output, g·km^{-1}, and year						
	1976	1977	1978	1979	1980	1981	1982
1970 Clean Air Act	9.4	2.1 →	→	→	→	→	→
July 1st, 1977	9.4 →	→	2.1				
1977 Amendments	9.4 →	→	→	→	4.4	2.1† →	→

*Based on data of the Council on Environmental Quality, 1978.
†Standard to be sustained for 31,000 km.

1956; Wright et al, 1975), but much higher readings (5 to 20 Pa) are observed in such sites as overhanging entrances, poorly ventilated tunnels, and underground garages (Ayres et al, 1973; Bithel, 1970; Conlee et al., 1967; Lawther et al., 1962; Miranda et al., 1967; Ramsey, 1967; Rispler and Gilbert, 1970; Wilkins, 1956; Wright et al., 1975).

The impact of automotive exhaust is greatest upon the pedestrian (Chaney, 1978; Clayton et al., 1960; Godin et al., 1972) and the cyclist (Kleiner and Spengler, 1976). The motorist who drives with the windows and ventilators of his vehicle closed gets some buffering from the cab, encountering CO pressures of perhaps 1 to 2 Pa, but with open windows, figures of 2 to 5 Pa are encountered (Brice and Roesler, 1966; Haagen-Smit, 1966). Carbon monoxide concentrations fall rapidly on moving away from the highway (Godin et al., 1972). Pollution also takes some time to penetrate homes and offices, and the sharper peaks of contamination are avoided (General Electric, 1972; Schaplowsky et al., 1974; Yocum et al., 1971). Unlike many air contaminants, there is little tendency for carbon monoxide to become adsorbed on walls, curtains, and furnishings, but indoor concentrations can exceed ambient levels, particularly if there is a local source of carbon monoxide such as cigarette smoke, heating and cooking appliances, or leakage from an attached garage (Benson et al.,. 1972; General Electric Company, 1972).

Industrial

A wide range of occupations involve a danger of carbon monoxide exposure (see Table V). At the First International Congress of Labour (1919), attention was called to the steadily increasing industrial production of carbon monoxide, and delegates called for governmental investigations that would lead to the prevention of carbon monoxide poisoning (K. H. Cooper, 1966). On five criteria—proportion of the working population exposed, relative toxicity of the gas, incidence of CO-related disease, quantity of substance manufactured, and likely trend of future emissions, a more recent panel of American Industrial Hygienists assigned top priority to the need for a clear industrial exposure standard for this gas (A. D. Hosey,

Table V
POTENTIAL OCCUPATIONAL EXPOSURES TO CARBON MONOXIDE*

(a) Exposure to coal and coke fires
Bakeries—boiler rooms—brick burning—cooks—firemen—furnace starters—glass blowers—heat treaters—locomotive drivers and stokers.
(b) Metal foundries and refining
Blast furnace—brass foundry—lime manufacture, reduction of metallic oxides—nickel refining and smelting—steel manufacture—welding.
(c) Exposure to gasoline exhaust
Vehicle operators—garage workers—traffic and tunnel police, vehicle inspectors—marine surveyors—longshoremen and warehousemen, forklift operators—street workers—aircraft pilots
(d) Exposure to manufactured gas
Gas vehicle operators—production of coal, water, and producer gas—petroleum industry—sewer repairs—coal tar distilling
(e) Exposure to chemical processes
Manufacture of acetic acid—acetylene—alcohols—ammonia—beer—methanol, organic chemicals—oxalic acid—paper manufacture—paint solvents—phthallic anhydride—soda—zinc oxide
(f) Miscellaneous
Graphic art—tending of livestock—space exploration—submarine operation—cigarette smoke

*Based in part on W.M. Gafafer, 1964 and National Safety Council, 1969.

1970). In 1977, United States industrial and related sources included forest fires (5.4×10^6 short tons of CO), agricultural fires (4.7×10^6 tons), solid waste disposal (2.9×10^6 tons), chemical manufacture (3.1×10^6 tons), petrol refining (2.6×10^6 tons), metal refining (2.2×10^6 tons), coal-refuse burning (0.6×10^6 tons), and structural fires (0.1×10^6 tons) (Environmental Protection Agency, 1979).

Exposure to Coal and Coke Fires

The hazards encountered by miners (Rodgers, 1926; Notov, 1959; Haldane and Priestley, 1935), drivers of steam locomotives, and men cooking food over charcoal fires have already been noted. The employees of bakeries, brick kilns, and glass works may all be at risk, along with those working in boiler rooms and furnaces (Gafafer, 1964).

Gas and charcoal cooking still causes appreciable CO exposure among grill workers in restaurants (Hosein and Bouhuys, 1979) and the slum populations of underdeveloped countries (Sofuluwe, 1968). During cold weather, the use of wood, coke, or gas fires inside small and tightly sealed buildings can also give rise to carbon monoxide poisoning (Wright et al., 1974; Kelley and Sophocleus, 1978).

Manz et al. (1976) found that, in contrast, the nonsmoking workers at a modern coke production facility showed no significant increase of carboxyhaemoglobin over the working day.

Major structural fires expose firemen to a combination of low concentrations of oxygen and high concentrations of carbon monoxide (Gordon and Rogers, 1969; Gregory et al., 1969; Goldsmith, 1970; Sammons and Coleman, 1974; Autian, 1976; Barnard et al., 1976; Levy et al., 1976; Loke et al., 1976; Petajan, 1976; Radford et al., 1976; Von Lupke and Schmidt, 1977; Terge et al., 1977; Griggs, 1977; Stewart et al., 1976; Terrill et al., 1978). Blood carboxyhaemoglobin readings of 10 to 14% have been observed in firefighters returning from a major blaze (Loke et al., 1976; Radford et al., 1976). Most city fire departments are now equipped with breathing apparatus (Shirling, 1961), but problems can arise from malfunction or exhaustion of the oxygen supply; moreover, in remote situations such as forest fires, breathing apparatus are not always available. It has been suggested that carbon monoxide exposure makes firemen vulnerable to disorders of cardiac rhythm (Gordon and Rogers, 1969) and other manifestations of ischaemic heart disease (Mastromatteo, 1959a, b; Barnard et al., 1976). Nevertheless, it remains hard to disentangle the relative importance of employment-related CO exposure, cigarette smoking, and long periods of inactivity between fires.

Metal Foundries and Refining

Since carbon and carbon monoxide both function as reducing agents, they are commonly used in the refining of oxide-containing ores to the pure metal. Grut (1949) noted significant CO exposure in casting, welding, the tending of furnaces, and related operations such as boiler cleaning, chimney sweeping, and the transport of hot coke or slag. A typical blast furnace yields about 3 kg of carbon monoxide for every kg of iron that is produced (G. M. Davies et al., 1965; Morel and Szemberg, 1971; Butt and Davies, 1974; Ejahn-Berdyer, 1973; Farmer and Crittenden, 1929; Hernberg et al., 1976; Jones and Walters, 1962; Lanino, 1964; Ministry of Labour, 1965; Virtamo and Tossaveinen, 1976; Pham et al., 1977; Seppänen and Uusitalo, 1977; Sedlak et al., 1978). The exhaust from the cupola contains 20 to 30% carbon monoxide, and if there is a leak this can seep into the working environment. In addition to formation during melting and casting, carbon monoxide is evolved by the oil burners used to heat the ladles and by internal combustion engine loaders. Air samples drawn from the immediate vicinity of the cupola have an average CO pressure as high as 24 Pa (Virtamo & Tossaveinen, 1976). At a modern Yugoslav steel plant, 2.8% of gas samples had a CO pressure of over 10 Pa, and a further 7.8% were between 5 and 10 Pa (Bauer et al., 1976). Other data from the literature include 11 Pa for the casting area of foundries, 8.5 Pa in the breathing zone for foundry workers, 2 Pa in the vicinity of electrical furnaces, and 2 Pa in the melting and casting areas of copper alloy foundries.

It has been known for many years that the blood carboxyhaemoglobin level of blast furnace workers is much above the population average (Farmer and Crittenden, 1929). In many British iron works, it was left to "the prevailing wind to blow the gas away from the workman. He was allowed to work until the effects of the gas were apparent, and was then replaced by a fresh workman while he retired to a clear atmosphere to recuperate . . . vomiting, taking aspirin for the headache, drinking copious drafts of beer to flush out the system, and then taking more aspirin to relieve the effects of the beer!" (J. G. Jones, 1961). Despite vigorous safety programmes, accidents remain common (Jones and Sinclair, 1975). In a single year, the reports for two British blast furnaces showed fifty-one cases of severe poisoning and one fatality (Jones and Walter, 1962). Nonsmoking blast furnace operators had blood carboxyhaemoglobin readings of 2.3 to 14.9% at the end of their shifts, the largest increase (10.0%) being seen in a gas main attendant. Moreover, when they reported for work the next morning, the nonsmokers still had an average blood carboxyhaemoglobin of 2 to 4%, suggesting that not all of the burden of carbon monoxide had been eliminated by a night's rest (Jones and Walter, 1962).

At one foundry, most CO readings were below 4.9 Pa, but occasional figures of up to 20 Pa were recorded during the emptying of a dust catcher, leakage of a valve, and demolition of a damaged wall (Davies et al., 1965).

Exposure to Gasoline Exhaust

CARS IN CONFINED SPACES. The operation of cars in narrow city streets, tunnels, underground parking garages, border inspection stations, and service stations exposes many workers to quite high concentrations of carbon monoxide (Balabaeva and Kalpaznov, 1974; Chovin, 1967; De Bruin, 1967; Goldstein et al., 1975; Göthe et al., 1969; Luederitz, 1971).

Wilson et al. (1926) suspected slight carbon monoxide poisoning when police patrolling the congested business section of Philadelphia complained of headache, nausea, and muscle weakness; symptoms were most frequent in areas of high traffic density and were relieved if the officers concerned were transferred to less-polluted parts of the city. Van Slyke measurements indicated blood carboxyhaemoglobin ($\overline{COHb}$) readings in the range 5 to 13%. Moureau (1964) examined 157 nonsmoking Parisian policemen. Unfortunately, his experiments were conducted over a windy five-day period, and CO concentrations were relatively low (0.5 to 1.4 Pa); 42% of the men showed a "definite" increase of carboxyhaemoglobin over their shift, but none of them reached what Moureau regarded as an acceptable ceiling (4.8% $\overline{COHb}$). Further analysis of the data (Chovin, 1967) suggested that the increase of blood CO content over a working shift (Δ_{CO}, mg·dl^{-1}) could be calculated quite simply from the ambient carbon monoxide (C_{CO},Pa):

$$\Delta_{CO} = 0.24\ (C_{CO}) - 0.07$$

De Bruin (1967) compared nonsmoking traffic policemen in Amsterdam and Rotterdam with control office workers. Three to four hours of street patrol led to only a small (0.25 to 0.28%) increase of COHb. Claude et al. (1978) found no increments in nonsmoking police. Data on Swedish police (Göthe et al., 1969; Sappanen and Uusitalo, 1977) were likewise strongly influenced by smoking habits; nonsmoking officers in Malmö and Örebro had average COHb readings of only 0.6% at the end of a working shift. In the larger city of Stockholm, corresponding readings were 1.1% for suburban and 1.3% for central area patrols. Certainly, any build-up of carbon monoxide upon city streets is quite localized (Bové and Siedenberg, 1970; Brice and Roessler, 1966; Georgii et al., 1967; Godin et al., 1972; Ramsey, 1966; Wilkins, 1956; Wright et al., 1975). Godin et al. (1972) observed figures of 3 to 4 Pa on busy routes. Concentrations can rise to 20 Pa in badly ventilated areas such as long underpasses (Ayres et al., 1973; Bithel, 1970; Conlee et al., 1967; Lawther et al., 1962; Miranda et al., 1967; Ramsey, 1966; Rispler and Gilbert, 1970; Wilkins, 1956; Wright et al., 1975) but drop to levels of less than 0.5 Pa on back streets, broad highways (Rose and Smith, 1962; Rose et al., 1965), and even sidewalks (Godin et al., 1972). Wright and Shephard (1978) found a substantial (1.2%) increase of COHb among smoking policemen over the afternoon shift, this being related fairly closely to their reported cigarette consumption. As a group, the nonsmokers showed no change of COHb in the morning and only a statistically insignificant 0.2% increment over the afternoon portion of their shift. However, occasional individuals showed larger increments. Two nonsmoking officers assigned to vehicular patrols in the downtown area of Toronto showed 1.1 and 3.8% gains of COHb, while a nonsmoker stationed in an underground car park had a reading of 3.5% COHb at the end of his day's work. Data for the nonsmoking officers conformed to the equation

$$\Delta\ \text{COHb}\%\ (\log_{10} X + 2) =$$
$$0.31 \times 0.11\ (\text{patrol hours, } \log_{10} X + 1) + 0.12\ (\text{City CO ppm } \log_{10} X + 1)$$

Figures for Calgary (where hazards are increased by cold, a higher altitude, and narrow streets) seem somewhat higher (N. J. Smith, 1976); readings for nonsmoking police averaged 2.5 to 3.5%, with peak values of 4.8% for a central area policeman and 7.0% for a commissionaire issuing parking tickets. The Calgary police apparently began work with a high COHb (2.3 to 2.4%) because duties were assigned in a building that included a "parkade" and an automobile repair shop.

In tunnels, repairs are usually carried out at night, when traffic volumes are light; the main risk is encountered by foot patrols on the day shift (Ayres and Evans, 1973). The ventilation system is such that CO concentrations are held below 20 Pa (Haldane and Priestley, 1935) or 40 Pa (United States Bureau of Mines, 1927); values greater than 25 Pa are still quite frequent (Miranda et al., 1967; Chovin and Truffert, 1973; Sterling and Kobayashi, 1977). Sievers et al.

(1942) noted that the average CO pressure in New York's Holland tunnel was 7 Pa with occasional peaks to 20 Pa; at the end of a two-hour shift, patrolmen had a COHb of 6%, but they showed no unusual incidence of cardiac disease or neurological disorders. The average CO reading for the most polluted of the Triborough tunnels, also in New York, was 6.3 Pa with a maximum hourly value of 21.7 Pa; 85% of the smokers and 47% of the nonsmokers employed by the bridge and tunnel authority had a COHb of over 3%, the average for the nonsmokers being 2.9% while working in the tunnel and 2.1% while working on the bridge (Ayres et al., 1973). Figures for a five-lane auto tunnel in Paris (Chovin and Truffert, 1973) were similar, with an average CO pressure of 6 Pa and peaks of over 20 Pa that persisted for several hours at busy holiday seasons. The accumulation of carbon monoxide is particularly serious if the tunnel is at high altitude (Miranda et al., 1967).

Modern parking garages generally have fair cross-ventilation, but in some older designs the only "fresh" air comes from the already polluted atmosphere of the street. Likewise, in modern underground garages, the CO pressure is less than 3 Pa (Wright et al., 1975), but in older structures inadequate ventilation subjects toll collectors and service station operators to a substantial eight-hour dose of carbon monoxide (Buchwald, 1969; Fristedt and Akesson, 1971; Trompei et al., 1964; Ramsey, 1967; N. J. Smith, 1975; Burgess et al., 1977). At one parking tower in Dayton, Ohio, CO concentrations ranged from 0.7 to 24 Pa (mean 5.9 Pa), the blood COHb of nonsmoking employees increasing from 1.5 to 7.3% over a typical shift (Ramsey, 1967); although there was no cumulative increase of morning COHb over the week, the haemoglobin levels of the parking attendants were on average 5% higher than in control subjects. Buchwald (1969a) reported even more alarming figures (COHb 18 to 22%) among service station employees at an underground parking garage. Further study showed average CO pressures of 3.3 Pa in ticket booths and 6.3 Pa in the automechanic area, but at both locations peak CO values exceeded 24 Pa (N. J. Smith, 1975). Employees complained of headaches and chronic fatigue, with end of shift COHb readings of 13.2% in a nonsmoker and 23.6% in a smoker: the haemoglobin concentration in the latter individual was over $19 g \cdot dl^{-1}$.

"Drive through" facilities expose workers to long lines of idling cars (Sievers et al., 1942; Myronuk, 1977; Burgess et al., 1977; Catcott and Xintaras, 1962; Cohen et al., 1971a; Johnson et al., 1974). In one mechanical fitness testing station, ambient CO pressures ranged from 1 to 14 Pa (average 5 Pa), COHb readings were 3.4% and 3.8% for nonsmokers and smokers respectively (Hofreuter et al., 1962), and carboxyhaemoglobin values were 4% higher than in office workers. Toll collectors on an expressway were normally exposed to an eight-hour average of 2.3 Pa carbon monoxide, but on one day an average of 5.7 Pa was seen (Xintaras et al., 1974). Customs and immigration officers on busy highways also encountered peak CO pressures of more than 10 Pa, 26% of samples exceeding 5 Pa (De-Bruin, 1967; Conlee et al., 1964). Detailed analysis showed a wide range of values (0.5 to 17 Pa, depending upon wind velocity and direction); eight-hour averages were greatest (11.4 Pa) from midnight to eight AM, when vertical air movement was minimal (S. I. Cohen et al., 1971a). Considering all shifts jointly, the COHb of nonsmokers increased from 1.5 to 3.6% over the working day, while corresponding figures for smokers were 4.8 and 6.4%.

It was suggested in one report that conditions could be improved by the installation of an electrically operated conveyor belt for the car, as at some car wash facilities.

The serious hazard presented by vehicle repair shops has been recognised for many years. Henderson and Haggard (1923) commented that "partial asphyxiation resulting in the typical carbon monoxide headache is a frequent, and in some cases, an almost daily occurrence with many of the employees in the better garages." Even worse conditions prevailed in smaller repair shops. At one establishment, nearly all employees went home with a headache every day, and at another company a substantial number of working days were lost from "headache and nausea." In some instances, the only provision for dispersal of exhaust fumes was natural diffusion; of thirty grab air samples, one showed a CO pressure of over 200 Pa, nine exceeded 150 Pa, and a further seven exceeded 100 Pa (Ciampolini, 1924). In contrast, Katz and Frevert (1928a) found that the average CO pressure for government garages did not exceed the permitted occupational standard of the era (10 Pa), while in commercial garages the highest readings were 43 Pa (one-hour average) and 16 Pa (twelve-hour average). It is possible that earlier investigators collected their grab samples at times of peak activity, while Katz and Frevert (1928a) biassed their data downwards by studying the large garages at a season when doors were wide open. The main factors influencing CO concentration were the number of vehicles being tested, their distance from the sampling port, and local ventilation. Bloomfield and Isbell (1928) demonstrated the importance of building size in a survey of twenty-seven repair shops. While the average CO pressure was 21 Pa, 18% of the readings exceeded 40 Pa, and in one garage a limited floor area, low ceiling, and poor ventilation were associated with a value of 110 Pa. Pilaar (1929) noted that when the doors of a small garage were open the COHb of the mechanics averaged 2.2%, but it rose to 7.2% when the doors were closed to conserve heat. Despite more stringent legislation, the problem persists. A survey of 268 garage and service station employees from Alberta showed occasional COHb readings as high as 16% after five hours of work (Buchwald, 1969b), although only 30% of nonsmokers had values higher than 5%. A report from Switzerland noted carboxyhaemoglobin levels of 3.3 to 6.6% in nonsmoking motor mechanics at the end of the working day, normal readings being restored with two days' absence from work (Anderhub et al., 1970).

Other occupational groups with potential exposure to high concentrations of carbon monoxide include street repair workers, street cleaners, drivers of city delivery trucks, cyclists, street vendors, and foot messengers (Kleiner and Spengler, 1976; Clayton et al., 1960; Chaney, 1978; Godin et al., 1972. A study of street cleaners found an average COHb of 3.3%, but values for the two nonsmokers were only 1.2% and 1.4% respectively (Gettler and Mattice, 1933). Following seven hours of work, the drivers of cabs, delivery vans, and hearses showed an average increase in COHb of 0.25% (De Bruin, 1967); however, bicycle riders showed high values (1.9%) both before and after work.In downtown Boston, cyclists making trips of ten- to forty-five-minute duration were typically exposed to carbon monoxide concentrations of 1.1 to 1.5 Pa; in no instance did the trip-averaged exposure exceed 3.5 Pa (Kleiner and Spengler, 1976). Likewise, the COHb after a ten-mile cycle journey in central London was reported as only 0.5% (R. Williams, 1979). Nonsmoking pedestrians in Union Square, New York had an average COHb of 1.94% (Ayres et al., 1973); joggers, incidentally, are at even greater risk because of a high respiratory minute volume (Finkel, 1972).

Vehicle drivers are themselves exposed to carbon monoxide from other traffic, leakage of their own vehicle's exhaust system, and any accumulations of cigarette smoke (Federal Highway Administration, 1977). The open car windows of summer assure almost instantaneous exposure to street concentrations of carbon monoxide (Brice and Roessler, 1966; Haagen-Smit, 1966). In winter, cab concentrations depend upon the CO content when parked, the cab volume, the location of the "fresh" air intake (normally close to the tail pipe of the vehicle ahead but sometimes on the roof), the capacity of the blower system, and additional effects of vehicle and wind speed. Chaney (1978) noted that when traffic on the Los Angeles freeway slowed to 16 $km \cdot h^{-1}$ the CO pressure rose to 1.5 Pa and if the vehicles were halted it increased further to 4.5 Pa; high readings were particularly likely when following an older vehicle at increased altitude. When tunnels are crowded, the CO pressure may rise to 40 Pa in cars and 80 Pa in buses (Sterling and Kobayashi, 1977). A faulty exhaust system can also induce COHb readings of 10 to 15% in nonsmoking drivers (Wright and Shephard, 1978). CO pressures of 8 to 9 Pa have been recorded when several heavy smokers have undertaken a long journey in a small car with closed windows (Harke and Peters, 1974; Harke et al., 1974). In a taxicab, smoking-induced CO maxima of 3 to 4 Pa are more likely, dropping to 0.5 to 0.6 Pa with a high wind speed and ventilation of the cab. N. J. Smith (1976) noted that nonsmoking taxi drivers in the city of Calgary had COHb readings of 2.5 to 3.5%, compared with only 2% in nonsmoking bus drivers operating in the same city.

FORKLIFT TRUCKS. Forklift trucks (Goldsmith, 1970; Petrov, 1968) are often operated in confined spaces such as ships (Breysse and Bovée, 1969; Gibbs et al., 1973), railway box-cars (D. A. Hall et al., 1957), trailers (Linch and Pfaff,

1971), and warehouses (Bloomfield, 1957; D. A. Hall et al., 1957; Linch and Pfaff, 1971). Critical factors governing the severity of exposure include not only the emission rate for the vehicle but also the volume of the operating space, the extent of natural or forced air currents, and the respiratory minute volume of the worker. The truck driver may inhale only one-third as much contaminated air as a busy stevedore in an adjacent recess.

Conditions are particularly adverse in cold stores because of a desire to minimise refrigeration costs. Bloomfield (1957) described the case of a man driving a propane-powered forklift in a store of 250 m^3 capacity. After he complained of dizziness and headache, air sampling revealed a CO pressure of 40 Pa; it was then decided to replace the truck with an electrically operated vehicle.

Linch and Pfaff (1971) studied a warehouse where a gasoline-powered forklift was being used to stack drums. The loading ramp doors and ventilators were both closed for warmth. In this situation, half of the grab samples showed a CO pressure of 6 to 10 Pa.

On board ship, a hold with air pockets seemed a greater hazard even than the use of canvas awnings during rainstorms (Breysse and Bovée, 1966). At the beginning of the shift only 3% of workers had a $\overline{\text{COHb}}$ greater than 10%, but by the end of the shift 10% of the sample exceeded this limit, forklift drivers being at greater risk than winch operators or stevedores. Longshoremen do not necessarily penetrate areas where the forklift is operating; average $\overline{\text{COHb}}$ readings for nonsmokers are 1.2% (Goldsmith, 1970) or 2.0% (Gibbs et al., 1973).

A similar type of problem is encountered in ice arenas, where petrol or paraffin-powered resurfacing equipment can expose the driver, the hockey players, and the public to carbon monoxide pressures of more than 5 Pa (D. E. Anderson, 1971; B. L. Johnson et al., 1975; Spengler et al., 1978).

Other vehicles. We have noted the use of producer gas-fuelled vehicles during World War II. The current energy crisis may revive interest in this type of equipment (Grut, 1949). Since the CO content of the gas is very high, any leakage into the cab of the vehicle is a serious risk to the driver. Workers in one producer gas-powered transport system showed a blood $\overline{\text{COHb}}$ of around 5% (although smokers were not distinguished). A variety of vague symptoms (headache, fatigue, irritability, loss of memory, confusion, dizziness, disturbances of mood, cardiac symptoms, digestive disturbances, loss of libido, excessive sweating, and thirst) were all attributed to possible chronic carbon monoxide poisoning (Grut, 1949), although critics pointed out that the men concerned all knew that they had been exposed to carbon monoxide (Lindgren, 1961).

Aircraft emissions are greatest for piston-operated machines such as light aircraft, short takeoff and landing (STOL) aircraft, and helicopters: typical percentages of carbon monoxide (by weight) are 8.8% of exhaust gases at takeoff, falling to 3.0% when cruising (Randel, 1971). In some light aircraft, the exhaust manifold runs along the side of the cabin wall, with the possibility of direct inhalation of carbon monoxide through an open canopy or indirect contamination of the cabin heating system by deterioration in the manifold (J. White, 1932; Speert, 1946; Slusher, 1964; Howlett and Shephard, 1973). The current generation of large passenger aircraft provides fifteen changes of cabin air per hour and this is sufficient to hold CO pressures to 0.4 to 0.5 Pa in the face of a substantial number of cigarette smokers (Federal Aviation Administration 1971). However, under foggy conditions, limited air movement and long delays on taxiways can cause the exhaust of other aircraft to be pumped into the cabin (Flight Safety Foundation, 1973; Lorang, 1978; Sega, 1978; WHO, 1979). Measurements taken at J. F. Kennedy Airport (Long Island, N. Y.) showed cabin CO pressures of 0.5 to 3.6 Pa were sustained for thirty-nine to seventy-seven minutes.

A further hazard to aircrew and passengers is an accumulation of carbon monoxide within the terminal building, due to idling taxis, buses, and planes. Measurements at Toronto International Airport in 1970 showed more than half of readings on the arrivals concourse exceeded 5 Pa, with a peak of 15 Pa on one evening when there was little breeze (Bithel, 1970; Rispler and Gilbert, 1970). Notices asking drivers to turn off their engines while standing and reduction of congestion by the opening of a second terminal reduced CO pressures to 1.0 to 2.5 Pa three years later. In this specific situation, taxi drivers, aircraft service personnel, and baggage handlers were at greater risk than aircrew. Baggage trucks were driven into the terminal, creating CO readings of up to 11 Pa in the baggage-handling area.

At some small arctic airports, gasoline-powered generators are used to start the aircraft; although located in an aft compartment, such generators are a potential source of CO exposure for aircrew (Howlett and Shephard, 1973). Combustion heaters may also be used to heat cabins (Wright et al., 1974), and carbon monoxide readings of up to 100 Pa can arise in helicopters as a result of the firing of weapons (Hody and Schane, 1962). Carboxyhaemoglobin values of 9 to 15% are seen in the victims of some aircraft crashes (Smith et al.,1976; Dominguez, 1962; Steinbridge and Goldbaum, 1966); in some cases, carbon monoxide poisoning and loss of judgement may have contributed to the crash (Howlett and Shephard, 1973), but more often the fire and inhalation of carbon monoxide occur in the first few minutes after impact (Mason, 1962).

Poisoning can occur in other closed vehicles (Schulte, 1964; Toliver and Morris, 1966; Ross, 1967), including large motorized sledges (Pugh, 1959), schnorkelling submarines (Miles, 1962), and space capsules (D. M. Davies, 1975; Bondi et al., 1978; Bondi and Schaefer, 1979). In submarines, the long-term build-up of carbon monoxide is kept below 1.5 to 2.0 Pa through the use of catalytic burners (Bondi et al., 1978; Wilson and Schaefer, 1979). In space vehicles, the main source of carbon monoxide is the normal body CO production of 0.4 $ml \cdot h^{-1}$ per occupant (Coburn et al., 1963; Luomanmaki, 1966; Lynch and Moede, 1972; Delivoria-Papadopoulos et al., 1974). Finally, petrol-driven pumps can contaminate the air supplied to equipment used for underwater exploration (Miles, 1962; Furgang, 1972).

Manufactured Gas

Workers engaged in the manufacture of illuminating and heating gas and employees of the petrochemical industry have a substantial risk of carbon monoxide poisoning. Men repairing sewers are also endangered because a leak from a fractured gas main may accumulate in a large sewer.

Chemical Processes

A variety of chemical processes have carbon monoxide as a by-product. Examples include the manufacture of acetic acid, acetylene, alcohols, ammonia, beer, organic chemicals, oxalic acid, paper, phthallic anhydride, soda, and zinc oxide.

A special category of risk is exposure to dichloromethane (CH_2Cl_2), a compound found in household aerosols and paint removers; the concentrations encountered by painters (500 to 1000 ppm CH_2Cl_2) produce $\overline{COHb}$ readings of up to 10% (Stewart et al., 1972; Miller and Engel, 1973; Kubic and Anders, 1974; Ratney and Wegman, 1974; Carlsson and Huttengren, 1975; Stewart and Hake, 1976; Barrowcliff, 1978; Benzon et al., 1978; Boudene et al., 1978; S. Laham et al., 1978; Stevenson et al., 1978; Barrowcliff and Knell 1979). Details of the mechanism whereby the body converts dichloromethane to carbon monoxide are still discussed. The best methods of preventing this particular type of CO poisoning are addition of an irritant to the solvent (thus warning of dangerous concentrations) and improved ventilation.

Miscellaneous

Other vulnerable occupational groups (see Table V) have been brought to light through $\overline{COHb}$ determinations in blood donors (Stewart et al., 1973b; Torbati and Har-Kedar, 1974). Values of over 2% are encountered in processors of metal, chemicals, stone, and glass; printers; welders (Rogowsky et al., 1978); electrical assembly workers; and graphic art workers. To this list may be added livestock producers, who encounter organic sources of carbon monoxide (National Center for Disease Control, 1973; K. J. Konham et al., 1977).

Cigarette Smoke

There have been diatribes against tobacco from the time of Ben Jonson's "Every Man Out of His Humour" and the better-known "A Counterblast to

Tobacco" by James I of England (1566-1625). The latter complained "Herein is not only a great vanity, but a great contempt of God's good gifts, that the sweetness of man's breath, being a good gift of God, should be wilfully corrupted by this stinking smoke." Goethe (1749-1832) was perhaps the first to suggest an asphyxial reaction, indicating also the risk to the nonsmoker: "Smokers pollute the air far and wide and asphyxiate every respectable individual who cannot smoke in self defense."

Given the slow.combustion process, a high carbon monoxide concentration in tobacco smoke is not surprising (Dudley, 1888; Gettler and Mattice, 1933). Wahl (1899) found CO pressures of 2.3 and 6.2 kPa in pipe and cigar smoke respectively. Factors influencing CO generation include the closeness of packing of the tobacco, the rate of smoking, the nature of the leaf and any additives, the depth of inhalation by the smoker, and (in cigarettes) any dilution of smoke by air vents (Armstrong, 1922; Shephard, 1982; Lundh et al., 1977).

Mumpower et al. (1962) used a gas chromatograph to establish the CO pressure in successive puffs of a king-sized cigarette; average values were 3.1 kPa in the third puff, 4.4 kPa in the eighth puff, and 5.0kPa in the thirteenth puff. There were no major differences between filter and nonfilter cigarettes, but the CO delivery during the initial puffs was greater for the short regular-sized cigarettes (Newsome and Keith, 1965; Robinson and Forbes, 1975). For the purpose of testing cigarettes it is assumed that the average smoker inhales one 35 ml puff per minute, over an interval of two seconds (Guillerm et al., 1972; Mumpower et al., 1962). Over a six-minute period of smoking this would be equivalent to an ambient exposure to 40 to 50 Pa of carbon monoxide. Factors modifying this estimate include a reduced nicotine and tar content (which is encouraging longer, deeper, and more frequent inhalations of the smoke), venting of cigarettes (which gives some protection to the smoker but increases the exposure of bystanders), and increased use of additives and reconstituted tobacco (Shephard, 1982; Sutton et al., 1978; Creighton and Lewis, 1978; Wright and Shephard, 1978; Russell et al., 1973a, b; Aronow et al., 1971; Cohen et al., 1971; Ashton and Watson, 1970).

The CO pressure in the mainstream smoke from cigars (5.3 to 11.1 kPa) is generally higher than that yielded by cigarettes (Hoffman and Wynder, 1972). The sidestream smoke encountered by the nonsmoker also contains 4.7 times as much carbon monoxide as mainstream smoke (Bridge and Corn, 1972; Hoegg, 1972).

From the viewpoint of the general population, it should be stressed that the average smoker absorbs far more carbon monoxide from tobacco products than from the urban atmosphere; each cigarette boosts COHb by up to 0.8% (Karnik and Coin, 1980), and in most instances atmospheric contamination does no more than impede elimination of tobacco-generated carbon monoxide. A careful smoking history is thus a necessary prelude to any investigation of supposed industrial carbon monoxide poisoning. Moreover, the heavy smoker is already partially acclimatized to carbon monoxide, and experimental studies of responses to low CO concentrations must therefore ensure that all test subjects are nonsmokers.

The tobacco-generated carbon monoxide that accumulates in closed spaces includes the entire content of sidestream smoke and that percentage of mainstream carbon monoxide (usually 40 to 50%) not absorbed in the smoker's mouth or lungs. Exposure of the nonsmoker depends upon the proximity of smokers, the amount and type of smoke they produce, the volume and ventilation of the room, and any local obstructions to ventilation (Steinfeld, 1972; Shephard, 1982; *British Medical Journal,* 1978).

The pattern of smoking influences both the composition of exhaled smoke and the proportion of sidestream smoke produced per cigarette (Boger, 1929). Mainstream smoke contained only 3 ml of carbon monoxide when small, intermittent puffs were taken, but with a fast and furious pattern of smoking up to 16 ml of carbon monoxide was generated. Sidestream smoke yielded 8 ml of carbon monoxide, rising to 12 ml if there was continuous suction upon the cigarette.

The high concentrations of carbon monoxide reached in small enclosed spaces such as a European passenger car (8 to 11 Pa; Srch, 1967; Harke et al., 1974; Harke and Peters, 1974) have been noted above. Wahl (1899) machine-smoked twenty cigars in a sealed room of 66m^3 capacity and observed a resultant CO pressure of 15.6 Pa. Ignoring possible absorption by the smoker, Baumberg (1923) calculated that ten people who smoked continuously could induce a CO pressure of 21 Pa in a 27m^3 room. Hoegg (1972) noted a peak of 7 Pa when twenty-four cigarettes were smoked simultaneously in a 25m^3 room; others have found similar readings when substantial numbers of cigarettes were smoked in small and poorly ventilated spaces (see Table III) (Harke, 1970; Harke et al., 1972; Russell et al., 1973; Hess, 1971; Lightfoot, 1972). Harke et al. (1974) observed readings of up to 11 Pa when nine cigarettes were smoked inside a small car, but ventilation quickly reduced CO pressures to 1 to 2 Pa; the smoking of two cigarettes induced a reading of 2.1 Pa, but normal ambient levels were regained within a few minutes of ceasing to smoke.

Much lower carbon monoxide concentrations (0.5 to 1.0 Pa) have been observed under real life conditions such as conference rooms and parties where there has been modest ventilation of the room (Wakeham, 1977a, b; Bonham, 1973; Schmeltz et al., 1975). Anderson and Dalhamn (1973) found that six air changes per hour were sufficient to keep the CO pressure in an 80m^3 room occupied by seven smokers to 0.45 Pa; likewise, Jones and Fagan (1974) estimated that with twenty-five smokers each consuming four cigarettes per hour, two air changes per hour would hold the CO pressure in an 85m^3 room below 5 Pa, while twenty air changes per hour would reduce it to 0.9 Pa.

It seems generally accepted that pollution is particularly bad in bars and restaurants because many people have become conditioned to smoke while they drink. Sebben et al. (1977) found CO pressures of 0.8 to 3.8 Pa, depending upon occupancy, ventilation, the use of grills and barbecues, and proximity to busy streets. On average, indoor concentrations rose from 0.87 Pa when the bars opened at 6 PM to a plateau of about 1.5 Pa between 7 PM and midnight. Over the same period, outdoor concentrations fell progressively, emphasizing that the major source of the pollution was cigarette smoke. Fischer et al. (1978) and Chappell and Parker (1977) have also noted significant accumulations of carbon monoxide in taverns and nightclubs. Unfortunately, the bartenders and waitresses who must work in this environment are engaged in quite heavy physical activity, thereby increasing their inhaled dose of carbon monoxide and other air contaminants.

The carbon monoxide concentrations on public service vehicles are usually relatively low, despite frequent complaints about annoyance from the accumulation of cigarette smoke on buses, trains, and aircraft (Federal Aviation Administration, 1971; Godin et al., 1972; Seiff, 1973; Sebben et al., 1977). On one intercity bus (Seiff, 1973) a blower system provided fifteen air changes per hour. Under the worst possible conditions (fully loaded, with 50% of the passengers smoking continuously),the carbon monoxide pressure was 3.3 Pa, compared with 0.7 Pa outside the vehicle. In a realisitic worst case (rear 20% of the coach occupied by individuals smoking 50% of the time), readings were 1.8 Pa inside and 1.3 Pa outside the bus. On aircraft, concentrations of carbon monoxide were even lower (Federal Aviation Administration, 1971); air sampled from near the mouths of seated passengers typically had a CO pressure of 0.2 Pa, with only one sample exceeding 0.5 Pa.

We may conclude that the carbon monoxide emitted from cigarettes is a potential health hazard, that it can be controlled by a high level of local ventilation, but that the cost of such additional ventilation is a serious consideration with rapid escalation of fuel prices (Shephard, 1982; Chappell and Parker, 1977; Wakeham, 1977a, b; Hugod et al., 1978b; Badre et al., 1978; Grimmar et al., 1977; Burns, 1977; Valentin et al., 1978). Little (1976) made a specific analysis of the 1975 recommendations of the American Society of Heating, Refrigeration, and Air-Conditioning Engineers and concluded that in the absence of separate smoking and nonsmoking areas, the proposed reduction in the ventilation of public buildings could bring cigarette-related accumulations of carbon monoxide to three times the proposed ambient air quality standard.

Chapter 2

DETERMINATION OF CARBON MONOXIDE EXPOSURE

Carbon monoxide concentrations can vary widely over a distance of one or two metres (Godin et al., 1972). Although relatively precise physical and chemical methods are now available to assess exposure at one specific location, for many epidemiological purposes it remains preferable to use man as an integrating device, judging total exposure from the accumulation of carbon monoxide in biological tissues. The elimination of carbon monoxide from the body is a relatively slow process (half-time three-four hours), so that measurements of venous or capillary blood carboxyhaemoglobin and even gas-phase estimates of carboxyhaemoglobin provide a convenient method of assessing the total carbon monoxide exposure over a working shift (N. J. Smith, 1977). The approach of biological sampling is particularly valuable when there are wide spatial or temporal fluctuations of carbon monoxide concentrations.

DETERMINATIONS OF CARBON MONOXIDE IN AMBIENT AIR

Choice of Methodology

Factors influencing the choice of analytic methodology include linearity of response, specificity versus interference from other air contaminants, simplicity of measurements, reliability and stability of calibration, sensitivity, accuracy relative to known gas concentrations, response time and sample volume, and (particularly in field applications) ruggedness, portability, and cost (Radford and Rodkey, 1973; Air Pollution Control Association, 1970). In view of the cumulative nature of carbon monoxide poisoning, continuously recording instruments are preferable to devices for spot sampling (see Table VI).

Chemical Detectors

Many early studies of carbon monoxide exposure are based upon rather

Table VI
BEST PERFORMANCE OF SELECTED METHODS FOR ANALYSIS OF CARBON MONOXIDE IN BLOOD*

	Blood Sample (ml)	Detectable difference ($ml \cdot dl^{-1}$)	Analysis time (min)	Coefficient of variation (%)
Van Slyke gasometric	1.0	0.03	15	6
Syringe gasometric	0.5	0.02	30	2-4
Spectrophotometric	0.1	0.08	10	—
"Co-oximeter"	0.4	0.10	3	—
Infrared	2.0	0.006	30	1.8
Thermal conductivity	1.0	0.001	30	2.0
(Liquid gas chromatography)	0.25	0.006	3	1.7
Flame ionization (Liquid gas chromatography)	0.1	0.002	20	1.8

*Based in part on Dahms and Horvath, 1974.

tedious and imprecise chemical techniques, such as length of stain measurements or colour matching. Reactions included the discoloration of palladium-molybdenum indicator tubes (Shepherd, 1947, 1955), the liberation of iodine from heated iodine pentoxide (Teague, 1920), or the reaction of CO with a mixture of iodine pentoxide, selenium dioxide, and sulphuric acid (Jacobs, 1949, 1960; Silverman and Gardner, 1965; Stewart and Stolman, 1961). Potassium palladosulphate changed from yellow to black and iodine pentoxide from colourless to bluish green, while a mixture of palladium sulphate and ammonium polybdate changed from yellow to green or blue. Since the carbon monoxide was used as a reducing agent in these various processes, there was an unfortunate lack of specificity; particularly in industry there was serious interference from other air contaminants with reducing properties (gases such as acetylene, hydrogen sulphide, or hydrogen). Sometimes, an absorbent chamber was added to remove interfering gases.

There has been a fair variety of electrochemical carbon monoxide detectors. Sjöstrand (1948a) and Linderholm (1965) both measured the heat produced when a steady flow of gas was passed over hopcalyte. Others have employed a galvanic cell to detect the iodine liberated from iodine pentoxide (DuBois et al., 1966; Haagen-Smit, 1966; Hersch, 1964) or have made photometric measurements based upon the release of halogen from a palladium halide (LeMoan, 1952; Allen and Root, 1955a), the molybdenum blue released from a compound of palladium chloride and phosphomolybdic acid (Polis, Berger, and Schrenk, 1964), the coloured complex formed from palladium and nitrose-β-naphthol (Fabre et al., 1952), or the mercury vapour liberated from mercuric oxide (McCullough et al., 1947; Beckman et al., 1948; Müller, 1954; Robbins et al., 1968; Palanos, 1972). One mercuric oxide method has been adapted for continuous gas analysis (Seiler and Junge, 1970). The main disadvantage is that other constituents of auto exhaust (hydrogen and hydrocarbons) also react with mercuric oxide.

A recent form of electrochemical detector is the easily portable piece of equipment known as the "Ecolyser" (Bay et al., 1972; Bay et al., 1974; Blurton and Bay, 1974; Yamate and Inoue, 1973). The basic principle of the machine is that a catalytically active platinum diffusion electrode converts carbon monoxide to CO_2; the latter is determined in terms of the associated production of hydrogen ions. The response of the system depends upon the rate of diffusion of carbon monoxide through the membrane plus the volume and buffering of the solution in which the carbon dioxide accumulates. The generated current is linearly related to the partial pressure of carbon monoxide over the normal operating range (0.1 to 10 Pa). Early reports were optimistic that interference from temperature, humidity, and reducing air contaminants would be minimal (Bay et al., 1972). Effects from ammonia, hydrogen sulphide, oxides of nitrogen, sulphur dioxide, and methane are small, but more recent experience has shown some response to the organic vapours encountered in bars and streets contaminated by auto exhaust (alcohol, acetylene, and ethylene); unfortunately 1 Pa acetylene yields an apparent CO reading of 11 Pa, while 1 Pa ethylene yields a CO reading of 0.25 Pa (First and Hinds, 1976). Nevertheless, results from the Ecolyser correlate closely with gas chromatographic data when analysing pure gas mixtures (Ayres, 1974), and use of the apparatus has substantially advanced our understanding of the local accumulation of carbon monoxide in buildings and in busy streets. One limitation of the equipment is a lack of instantaneous response (90% response time, 10 seconds, delay time ~ 20 seconds, Bay et al., 1972). However, as we shall see, this interval of time is short relative to the usual biological response to carbon monoxide exposure.

Physical Detectors

In many industries, the need is for continuous rather than spot monitoring of carbon monoxide pressures—indeed, the appearance of increasing amounts of carbon monoxide in a mine ventilating shaft can give early warning of a fire as well as a respiratory hazard. Continuous monitors are usually based upon physical principles (Strange et al., 1960). The equipment is typically cumbersome, requires a stabilized A C voltage source, and is vulnerable to vibration; it is thus best operated at a fixed sampling site.

Nondispersive infrared analysers provide the primary reference method (Air Pollution Control Association, 1970; Pierce and Collins, 1971; Hill and Powell, 1968; Jacobs, 1960; Feldstein, 1967). The basic principle of such apparatus is that simple, heteratomic gases like carbon monoxide, carbon dioxide, and water vapour each have a characteristic absorption spectrum. Energy from an infrared source is split into two equal beams, both made intermittent by a rotating "chopper" blade (see Figure 1). One beam passes through a column of test gas and the other through a column of "clean" air; both beams then impinge upon the sensor cells, filled with the gas of interest and separated by a thin diaphragm. Since some infrared energy is absorbed in the analysis column, less residual energy reaches the sensor on the analytic side of the apparatus. A pulsating pressure signal, proportional to the concentration of the heteratomic gas, is generated. This is readily detected and amplified by standard circuits (Hill and Powell, 1968). With carbon monoxide, a long path-length or high pressure analysis is necessary to give the apparatus sufficient sensitivity (Hill and Powell, 1968; Jacobs et al., 1959); consequent disadvantages are a slow response speed and a need for a large sample volume. The instability that plagued early designs of infrared analyser has been largely overcome by split-beam operation, automatic zero and span adjustments, and a thermostat to control the infrared source (see Table VII). The absorption bands for carbon monoxide are centered about a wavelength 4.6 to 4.7μm. The response to increasing concentrations of CO is linear until the central absorption bands are saturated. Thereafter, energy is taken up by more peripheral bands, in a manner proportional to the square root of column length. A short column avoids saturation, but the system then has a poor signal-to-noise ratio, becoming vulnerable to dirt and fogging of the windows. The cell length is thus adjusted so that there is about a 20% attenuation of the signal when the maximum intended CO concentration is reached (10% alinearity). This residual alinearity can be corrected by a sequence of "linearizing" circuits. Interference arises from other heteratomic gases, particularly carbon dioxide and water vapour but including also such possible industrial contaminants as cyanogen, methane, diazomethane, nitrosyl chloride, ammonia, acetone, ethyl alcohol, acetylene, and propane (Berlin and Smeets, 1974). The water vapour can be removed by drying agents or cooling, but if this precaution is not taken or the drying agents become saturated, a substantial false CO signal will occur. The effect of carbon dioxide and water vapour can also be countered by passing the infrared beams through "optical filters" containing these gases; nevertheless, there is overlap between the absorption bands for carbon monoxide, CO_2 and water vapour at 4.6μm, and screening may be incomplete (Coburn et al., 1964: Pierce and Collins, 1971). When examining respired gas mixtures, account must also be taken of "collision broadening" of the infrared spectrum as the proportion of oxygen in the gas mixture changes (Hill and Powell, 1968). Industrial and environmental monitors incorporate a particulate filter to protect the analyser against solids that can scatter the infrared beam, but care must be taken that clogging of the filter does not create a significant pressure drop in the detector column, with falsely low carbon monoxide readings. Periodic checks of the system for such problems as leakage are also necessary (Environmental Protection Agency, 1975). The data obtained remain somewhat vulnerable to details of technique.

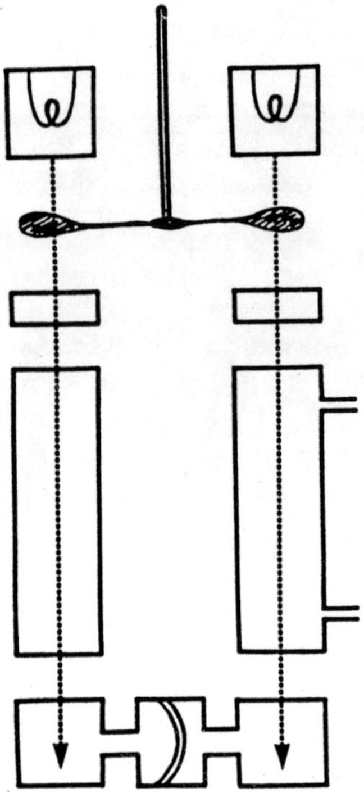

Figure 1. An illustration of the principle of infrared carbon monoxide analysis.

Table VII
PERFORMANCE SPECIFICATIONS OF U.S. ENVIRONMENTAL PROTECTION AGENCY FOR NONDISPERSIVE INFRARED ANALYSERS*

Range	0-5 Pa
Noise	$<$0.05 Pa
Lowest Detectable	CO 0.1 Pa
Interference	
Individual substances	$<$0.1 Pa CO equivalent
Total	$<$0.15 Pa CO equivalent
Zero drift	$<$0.1 Pa in 24 hours
Span drift	$<$0.1 Pa in 10 minutes
Lag time	10 minutes
95% response time	$<$5 minutes
Precision	$<$0.1 Pa

**Federal Register*, 1973.

In the United States, for example, a change from mono- to twin-beam detectors and the substitution of air/CO for N_2/CO calibration mixtures led to a substantial drop in the reported CO concentrations (Tiao et al., 1975). Because of possible absorption of CO on cylinder walls (Brief et al., 1967), the optimum method of calibration is to mix pure CO and CO-free air, using a precision dilution pump.

Possible developments of infrared technology include the use of a scanning, second-derivative spectrometer (currently available for ultraviolet wavelengths, Lawrence Berkeley Laboratory, 1973), resolution of individual absorption lines by Fourier transform spectroscopy (Bell, 1972; Hanst et al., 1973), and gas filter correlation spectroscopy (Burch et al., 1976). In the last approach, a collimated infrared beam is passed alternately through a carbon monoxide and a neutral filter, giving rise to a difference signal when the beam reaches the column containing the unknown concentration of carbon monoxide. This approach eliminates all response to carbon dioxide and greatly minimizes the effects of water vapour. A variant ("dual isotope fluorescence") illuminates the sampling columns alternately with infrared beams characteristic of normal carbon monoxide (CO^{16}) and its isotope (CO^{18}); there is thus a cyclic variation in the transmitted energy proportional to the amount of CO^{16} that is present; interference from carbon dioxide and water vapour is effectively eliminated (McClatchie et al., 1972).

Some authors have adapted gas chromatography to the analysis of carbon monoxide. Columns have been packed with silica gel, a mixture of silica gel and iodine pentoxide, zeolite, or a molecular sieve (Szulcewski and Higuchi, 1957; Smith, Swineheart, and Lesnini, 1958; Kyryacos and Boord, 1957; Forster, 1969a; Commins, 1957). Low concentrations of carbon monoxide can be detected by the use of capillary columns, with reduction to methane and detection by flame ionization (Ettre, 1961; Porter and Volman, 1962; Rodkey, 1970). Problems with this type of instrument include consistency in the pattern of sample injection, stability of carrier gas flow, and constancy of oven and detector temperatures. The equipment is very sensitive and usually avoids problems from interfering gases, but it is much more difficult to operate than is a nondispersive infrared gas analyzer; if there is a rôle for gas chromatography, it is probably in the analysis of the small amounts of gas liberated from blood specimens rather than in the determination of ambient gas concentrations (Maisels et al., 1971; Rodkey, 1970).

Although mass spectrometers are now widely used by respiratory physiologists, there are difficulties in applying this technique to the analysis of carbon monoxide (Forster, 1969a). Expensive modifications are necessary to distinguish carbon monoxide from the more prevalent nitrogen (which has the same molecular mass). Carbon monoxide also tends to be produced in the ion-generating section of the apparatus. Finally, the machine operates in a near-vacuum. It is thus very suitable for determining gases that are present in high concentrations (nitrogen, oxygen, and carbon dioxide) but becomes inaccurate when detecting gases such as carbon monoxide, which are usually present in quite low concentrations.

Sampling Networks

Carbon monoxide concentrations vary widely with respect to time and space whether monitoring urban or industrial exposure. Considerable thought must thus be given to design of an appropriate sampling network and to an appropriate averaging of the data (Gilmore and Hanna, 1976; Aron and Aron, 1978). Variables within the city include vehicle density and speed, windspeed and atmospheric stability, local air currents created by tall buildings, and local air pockets under bridges and partially enclosed structures. In industry, accumulations of carbon monoxide may be equally localized to specific sites at particular phases of a manufacturing operation.

The usual arrangement for urban sampling is a limited network of long-column twin-beam nondispersive infrared analyzers, placed at fixed sampling sites. Inlet probes are mounted at least three metres above the ground to avoid vandalism (Altshuler et al., 1966) and at least one metre from adjacent structures to avoid frictional and eddy distortions of the airstream (Nicholson, 1975). The United States Environmental Protection Agency (1977) recommended a minimum of one sampling station in a city of 100,000 population, with an additional 0.15 stations for every further 100,000 inhabitants. By 1977, reports were being received from 456 sites across the United States (Environmental Protection Agency, 1978).

The adequacy of such data depends on the proximity of the probe to major traffic routes. While microscale measurements are sometimes useful in gauging local compliance with air quality standards or assessing risks to specific occupational groups such as taxi drivers (Brice and Roesler, 1966; Godin et al., 1972; McCormick and Xintaras, 1962; Wilkins, 1956; Wright et al., 1975), larger scale measurements are essential to gauging regional pollution trends and assessing the diffusion of carbon monoxide from urban to rural areas. For macroscale readings, the sampling inlet must be kept away from local traffic sources; the pressure then recorded (0.3 to 1.0 Pa) reflects the general urban background (Colucci and Begeman, 1969; Hamming et al., 1960; Moureau, 1964; Ott and Eliassen, 1973).

When studying a suspected industrial problem, it may be sufficient to show very high carbon monoxide readings in a few grab samples. The traditional approach has been to use silica (length of stain) detector tubes, a small volume (100 ml) of air being drawn through the tube by a hand pump; the main disadvantage of such measurements is their limited accuracy (± 20%). More recently, the Ecolyser has been carried to supposedly polluted areas of a factory. It has a much greater sensitivity than the stain detector, although care must be taken to avoid false positive readings from organic vapours.

If an industrial process is well standardized, the best approach is to place a continuously recording infrared detector at the main exposure sites. Again, interference may arise from water and organic vapours, and appropriate optical filters must be provided. The instrument must also be regularly calibrated and well maintained, with backup equipment available if prolonged repairs are necessary.

When employees move between zones of differing risk, personal exposure must be assessed. The most common basis for such testing is to measure the accumulation of carboxyhaemoglobin in the blood. However, it is also possible for the worker to carry a battery-operated pump that samples air from around the mouth (S. A. Hall, 1973).

Timing of Samples

Exposure to a toxic material is commonly expressed as a concentration-time product (CT). This approach has some attraction when dealing with a substance such as carbon monoxide that can accumulate in the body. Nevertheless, the response to a sharp peak of pollution is not necessarily the same as that observed when a similar CT is reached by a long, low-intensity pattern of exposure; high concentrations of carbon monoxide can travel around the circulation as a "slug," penetrating to poison extravascular enzyme systems. It may thus be useful to report both short-term peaks and longer average CO concentrations. The blood reaches equilibrium with carbon monoxide over an eight-hour period (see Tables V and VI) so that it is not possible to calculate a realistic CT for an entire work shift. Possible alternatives are to specify an ambient concentration that should not be exceeded and to specify a blood carboxyhaemoglobin level that should not be exceeded.

Urban monitors commonly indicate morning and evening rush-hour peaks, with a sustained evening rise due to the usual increase of atmospheric stability after sunset (Brief et al., 1960; Colucci and Begeman, 1969; Godin et al., 1972). Superimposed upon this diurnal rhythm are seasonal variations in CO production and atmospheric stability plus local geographic factors that influence the development of thermal inversions (Shephard, 1974).

Urban pollution readings are generally reported as hourly, eight-hourly, twenty-four-hourly, monthly, or seasonal averages. A case can be made for shorter averaging periods for high-risk situations such as tunnels, underpasses, and narrow urban canyons. Nevertheless, the volume of data generated by permanent monitoring stations is too large to permit the wide application of this approach, and the relatively slow fluctuation of the urban "background" of carbon monoxide does not justify shortening the averaging period from sixty to fifteen minutes. The arithmetic mean and peak values are heavily influenced by occasional instrumental aberrations, and it has been argued that the median or the geometric mean are more appropriate statistics for what is often a log-normal distribution. Air quality may also be reported as a moving average rather than data for a fixed time span such as 24.00 to 8.00 hours (McMullen, 1975). Statements may be included on the highest or the second-highest figure reached over a specified period such as a month or a year. Air pollution experts point to the fatal London smogs of 1952 and 1956, arguing that the peaks of pollution are the cause of fatalities. The second-highest value is thus suggested as a safety limit that should not be exceeded on a significant number of occasions in a given year. The argument is probably valid for fatal peaks, but is more difficult to extend to the usual responses induced by exposure to lower concentrations of carbon monoxide. If exposure to the pollutant is increasing the risk of traffic accidents, for example, emergencies are probably occurring on a random basis, and the noting of a single peak of pollution does not indicate all motorists who are at risk. In such a situation, a measure of central tendency gives more health information than the reporting of an occasional outlying aberrant result. A possible compromise is to report both the median or the geometric mean *and* the number of occasions per year when a specified safety limit is exceeded. As a general guide, the eight-hour second-highest reading is about eight times the arithmetic mean for the year, while the one hour second-highest reading is about twelve times the arithmetic mean for the year.

Missing data can be estimated from a frequency distribution of the available carbon monoxide readings. However, statistical treatment of a sequence of measurements is complicated by autocorrelations of the data that reflect diurnal and longer cycles of meteorological conditions.

Industrial exposures are commonly reported as an eight-hour time-weighted average. This still reflects the risk encountered by many employees, but the situation is being influenced increasingly by adoption of the flexible working week, where individual shifts may have a duration of ten to twelve hours. The longer working day increases the final blood carboxyhaemoglobin level for a given carbon monoxide pressure, particularly if the work is sedentary in nature.

A case can also be made for reporting industrial exposures over shorter intervals (ten to 60 minutes), since very dangerous peak readings can become hidden within an eight-hour time-weighted average. Unlike urban exposure data, temporally related industrial measurements do not necessarily show autocorrelation (Hirtzel and Quon, 1979). Carbon monoxide readings may fluctuate widely within a shift, from shift to shift, from place to place within a factory, with season, and (in some industries) with the cycle of operations. A mathematician might argue that the chance of a worker sustaining an adverse reaction to a given episode of pollution is a statistical phenomenon, best reported by a measure of central tendency such as the median or the geometric mean. However, the complexities of union negotiations and governmental enforcement have encouraged acceptance of a threshold limit value (TLV), which cannot legally be exceeded.

Because of the log-normal distribution of readings, the mode is usually less than the mean, and a parsimony of data collection can lead to a significant underestimate of exposures. For a given averaging time S_1 the geometric mean

$\overline{C_g}$ is related to the arithmetic mean $\overline{C}$ by a function of the type

$$\overline{C_g} = a\overline{C} + b$$

where a and b are constants specific to a given averaging time (Inoue and Watanabe, 1977).

In general, the worst-case data should be sought. If seasonal factors are important (for example, the inefficient operation of cars and reduced ventilation of garages in cold weather or the reduced ventilation of taverns in summer heat), the data reported should describe conditions encountered during the most adverse season of the year. When discussing cyclic industrial processes or exposure to auto exhaust (with diurnal fluctuations of traffic density and atmospheric stability), the reported figures should refer to the worst eight-hour shift of the day. If there are large spatial variations of pollution within a factory, the reported statistics should describe the worst site occupied regularly by an employee over the working day.

DETERMINATION OF BLOOD CARBOXYHAEMOGLOBIN

Blood-Phase Determinations

Although the blood carboxyhaemoglobin level is commonly used to assess industrial exposures to carbon monoxide, it is important to exclude the possibility that high readings may have arisen from exposure to vehicle exhaust during the journey to work or from cigarette smoking. Policy decisions should be based on samples collected on arrival as well as the end of a working shift, and data for smoking and nonsmoking workers should be distinguished. A high initial reading may reflect either a polluted journey to work or incomplete elimination of carbon monoxide from the previous day's industrial exposure. If a high reading in a nonsmoker does not diminish over the working day, it may be presumed that the work place is polluted, although with some techniques it is important to exclude interference from other pigments, such as methaemoglobin and nitrohaemoglobin (Kerr, 1946).

Carboxyhaemoglobin standards are prepared by mixing known volumes of carbon monoxide-free and carbon monoxide-saturated blood. A further check on methodology is provided by including in the batch of blood specimens samples from nonsmoking subjects living in a rural area and having no occupational exposure to carbon monoxide.

Since there is a negligible arteriovenous carbon monoxide pressure gradient, it is immaterial whether arterial, capillary, or venous blood samples are submitted for analysis. Some early investigators adopted the expedient of shaking specimens of contaminated air with venous blood as a means of determining atmospheric carbon monoxide concentrations (Haldane and Priestley, 1935). The resultant cherry red solution of carboxyhaemoglobin was matched carefully with one of a series of standard carmine-tinted solutions.

Unfortunately, the carboxyhaemoglobin tended to dissociate in bright sunlight (Roughton, 1963; Haldane, 1896). The carmine standards were also somewhat unstable, and the match that was obtained depended upon the proportions of natural and artificial light illuminating the tubes (Haldane, 1895b). The carboxyhaemoglobin reaction was further influenced by intersubject differences of haemoglobin level (Haldane and Smith, 1898) and by room temperature (Hartridge, 1912b). It was soon agreed that the carmine method was inaccurate, except perhaps in Haldane's hands, and it may even have led him to erroneous views on such matters as acclimatization to carbon monoxide poisoning (Grut, 1949; Hartridge, 1912a).

The choice of analytic method now lies between liberation of the carbon monoxide and its analysis in situ by spectrophotometry. A variety of acids (sulphuric, lactic, hydrochloric, acetic, and phosphoric) are all capable of freeing carbon monoxide from haemoglobin, the process being helped forwards by oxidizing agents such as ferricyanide or biiodate of potassium. Thus, the classical Van Slyke manometric technique uses an acid ferricyanide solution to release carbon monoxide along with oxygen, carbon dioxide, and nitrogen (Van Slyke and Neill, 1934; Van Slyke et al., 1946). The carbon monoxide can be estimated as a residual gas pressure (Van Slyke and Neill, 1934) or as a pressure or volume change consequent upon chemical absorption of the carbon monoxide in cuprous chloride solution (Sendroy and Liu, 1930). However, such methods of analysis are lengthy (Gettler and Mattice, 1933) and require skilled personnel. Moreover, until the introduction of microgasometric procedures (Allen and Root, 1955b; Linderholm, 1965; Linderholm et al., 1966; Horvath and Roughton, 1942; R. G. Smith, 1975; Roughton and Root, 1945; Scholander and Roughton, 1942, 1943), substantial samples of blood were required.

While many investigators continue to extract carbon monoxide from the blood by means of the Van Slyke mercury pump, the trend for both macro- and microanalysis is to rely upon a chemical (Stetter and Blurton, 1976; Ammundsen and Grut, 1943; Christman and Randall, 1933; Lambert et al., 1972; Lee et al., 1975; Rice, 1952; Wennesland, 1940) or a physical (Coburn et al., 1964; Gaensler et al., 1957; Lawther and Apthorp, 1955; Moureau et al., 1957; Commins, 1975; Werner, 1976) rather than a volumetric or a manometric determination of the gas that is liberated. In some instances, gas chromatography has been coupled to the manometric extraction of carbon monoxide (Ayres et al., 1966; Collison et al., 1968; McCredie and Jose, 1967). Reports suggest the accuracy of these newer methods is at least comparable to that of the classical Van Slyke manometric procedure; indeed, one technique of vortex extraction allows 100 μl samples of blood to be analyzed with a coefficient of variation of 1.7% (Dahms and Horvath, 1974). Nevertheless, measurements remain time-consuming and demand considerable experience; many of the available procedures also use substantial samples of blood ($>$ 5ml).

Conway (1950) proposed an alternative chemical procedure, whereby sulphuric acid was used to liberate the carbon monoxide from haemoglobin within a microdiffusion cell. The gas was then absorbed by palladium chloride, with a release of hydrogen ions that could be titrated by microburette. Although considerable accuracy was claimed for the method it was tedious, and each analysis required a 2 ml specimen of blood.

Biological estimates are based upon differences in the absorption spectra of haemoglobin, oxyhaemoglobin, and carboxyhaemoglobin. Hufner introduced a single-beam spectroscopic method in 1889 (Hartridge, 1912a), and Hartridge (1912a, b) developed the twin-beam reversion spectroscope. In the latter instrument two spectra are viewed, one above the other, and one having the colours reversed; the distance separating the absorption bands in the upper and lower spectra varies with the COHb concentration. The method is well suited to its original application (fatal episodes of carbon monoxide poisoning) but lacks sensitivity at carboxyhaemoglobin readings of less than 35% (Douglas et al., 1962; Commins, 1975).

More recently, spectrophotometric methods have been applied to quite small samples of blood (10 μl to 0.5 ml). Linderholm et al. (1966), Blackmore (1970), Forster (1969), and the Air Pollution Control Association (1970) all concluded that such methods lacked the sensitivity needed for monitoring the carboxyhaemoglobin levels reached in ordinary urban exposures, but Lily et al. (1972) and Baretta et al. (1978) have subsequently claimed a standard deviation of only 0.13% for repeat determinations at carboxyhaemoglobin levels of less than 3%. One spectrophotometric approach treats haemolyzed blood with a reducing agent such as sodium thiosulphite. The relative proportions of haemoglobin and carboxyhaemoglobin can then be determined from the relative absorption of light at two appropriate wavelengths (Harper, 1952; Paul and Theorell, 1942). Klendshøj et al. (1950) claimed a good correlation with Van Slyke results when using data collected at 480 nm (where the spectra of reduced haemoglobin and carboxyhaemoglobin differ) and 555 nm (an isosbestic point, where absorption is equal for the two pigments).

A second possibility is to choose two wavelengths (530.6 nm and 583 nm) where the absorption of reduced and oxygenated haemoglobin is equal, but effects due to carboxyhaemoglobin differ (Ramieri et al., 1974).

Other alternatives to reduction of the blood sample are (1) to examine the optical density of the specimen before and after full oxygenation (Commins and Lawther, 1965; Lily et al., 1972), (2) to exploit the slow conversion of carboxyhaemoglobin to methaemoglobin (Paul and Theorell, 1942), or (3) to base determinations upon the less ready heat precipitation of carboxyhaemoglobin relative to oxyhaemoglobin (Whitehead and Worthington, 1961).

By coupling an absorption spectrophotometer to an analogue computer, measurements of carboxyhaemoglobin, oxyhaemoglobin, and haemoglobin concentration can be made on a single blood specimen (Maas et al., 1970;

Malenfant et al., 1968). Absorbance is measured at three wavelengths (in the commercially available "Co-oximeter," at 548, 568, and 578 nm), and analogue circuits then solve three simultaneous equations. Another procedure makes measurements at four wavelengths in the Soret band of the near ultraviolet spectrum (wavelength 390 to 440 nm, Small et al., 1971), where the separation of the oxyhaemoglobin and carboxyhaemoglobin absorption spectra is somewhat greater (see Figure 2). Under normal conditions of analysis, the system obeys the Beer-Lambert law for dilute solutions, and the absorbance A at any given wavelength is given by

$$A = Kl\,(a_{O_2}[1 - x - y] + a_{CO}[x] + a_m[y])$$

where K is a constant (the molar concentration of haemoglobin), l is the path-length of the cuvette, [x] is the fraction of carboxyhaemoglobin, [y] is the fraction of methaemoglobin, and [1 - x - y] is the fraction of oxyhaemoglobin. Measurements are made at 392, 413.2, 421, and 431.2 nm, yielding four simultaneous equations; a simple computer programme allows translation of the absorption coefficients into estimates of [x], [y], and [1 - x - y]. Limitations of the analysis include effects from temperature change, dirty cuvettes, and wavelength drift in the instrument, along with conversion of methaemoglobin to haematin, and loss of carbon monoxide from the blood as it is diluted. The blood sample required for the procedure of Small et al. is 100 μl, and an accuracy of ± 0.6% carboxyhaemoglobin is claimed up to a haemoglobin level of 25%.

Spectrophotometric determinations have an advantage over most chemical procedures in that relatively little blood is required. Moreover, the value reported is a carboxyhaemoglobin percentage rather than a carbon monoxide content. When converting chemical determinations to carboxyhaemoglobin percentages, it is necessary to obtain an accurate haemoglobin reading and to assume a fixed stoichiometric relationship between carbon monoxide and haemoglobin; while such an assumption is probably warranted for well-mixed blood, it has been disputed. Errors can also arise from failure to allow sufficient time for the conversion of carboxyhaemoglobin to cyanmethaemoglobin, the standard pigment used in haemoglobin determinations (Rodkey, 1967). Against these problems must be set one major difficulty of many spectrophotometric methods, namely interference arising from other pigments such as myoglobin, methaemoglobin, and bile salts. Fortunately, this complication is largely overcome in the four-wavelength ultraviolet technique of Small et al. (1971), and indeed methaemoglobin concentrations can themselves be determined with an accuracy of ± 2%.

Sawicki and Gibson (1979) have recently suggested exploiting the instability of carboxyhaemoglobin in bright light to obtain a very specific determination of blood carbon monoxide levels. Absorption is measured at 432 nm before

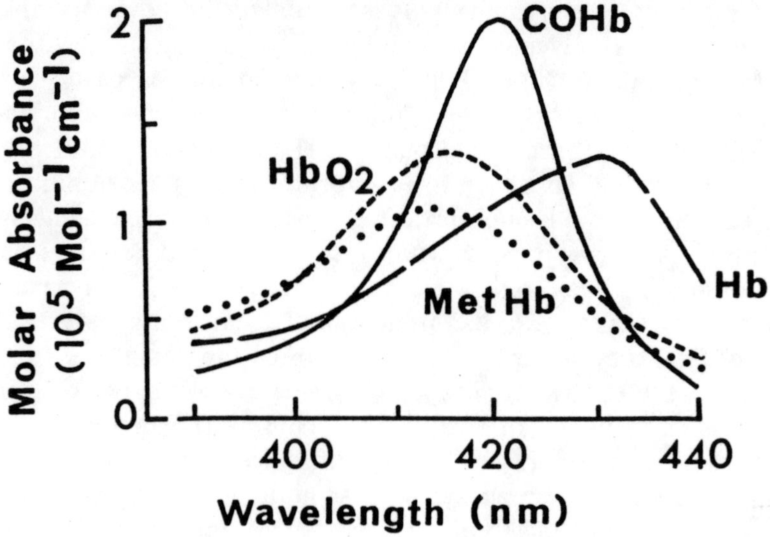

Figure 2. Absorption spectra in the Soret region of ultraviolet light. Data for oxyhaemoglobin, carboxyhaemoglobin, methaemoglobin, and reduced haemoglobin. Based upon results of Small et al., 1971.

and after photolysis of the carboxyhaemoglobin by a bright flash of light. The analysis requires only 5 μl of blood, and a resolution of better than 0. 1% saturation is claimed by the authors.

Bloodless Determinations

Venipuncture is unpopular with volunteers, and for this reason bloodless methods of determining carboxyhaemoglobin have become quite commonly used, particularly for mass sampling in urban and industrial surveys. Most procedures are based upon a rearrangement of Haldane's equation for the percentage of carboxyhaemoglobin

$$CO\overline{Hb}\% = HbO_2\%\, M\frac{P_{CO}}{P_{O_2}}$$

where $HbO_2\%$ is the oxygen saturation of the blood, M is a constant, usually in the range 195 to 250, and P_{CO} and P_{O_2} are the partial pressures of carbon monoxide and oxygen respectively.

One approach requires the subject to rebreathe for two minutes from a bag containing several litres of pure oxygen (Dahlström, 1955; Henderson and Apthorp, 1960; Sjöstrand, 1948a, b). A canister of soda lime serves to absorb carbon dioxide. At the end of the two-minute interval, the partial pressures of carbon monoxide and oxygen in the rebreathing bag are determined by standard methods of gas analysis. It is assumed that equilibrium has been established between alveolar and pulmonary gas, and since the oxygen concentration remains high, HbO_2 is close to 100%. Criticisms of the procedure include (1) incomplete equilibration between gas and blood phases (since the rate of diffusion is proportional to pressure gradient, there is inevitably a slow final approach to equilibrium values), (2) nonuniformity of alveolar carbon monoxide concentrations (this problem is particularly marked in individuals suffering from chronic obstructure pulmonary disease), (3) accumulation of carbon dioxide in the system (Grut, 1949; this can modify the oxygen dissociation curve of the blood and also causes hyperventilation), (4) changes of gas composition within the rebreathing bag due to diffusion, absorption in the rubber or plastic, and reactions with the soda lime (such difficulties are largely overcome by use of an aluminized mylar bag, with rapid analysis of samples), and (5) individual differences in the value of M. The last criticism is perhaps the most serious. Although some authors have found almost constant values of M, Dahlström (1955) noted figures ranging from 228 to 413. Interestingly, in his experiments M was more consistent (218 to 229) for the formation of an additional quantity of carboxyhaemoglobin than for determination of what was admittedly a relatively low baseline value. Carlsten et al. (1954) suggested the standard deviation of M was as much as ± 45. Part of this large variation probably reflects a failure to standardize conditions of measurement. Coburn et al. (1965) and Rodkey et al. (1974) both found M to range between 220 and 250 at physiological pH and body temperature. Roughton (1970) determined an average value of 246 but stressed that this varied over different segments of the oxygen dissociation curve.

Despite the theoretical problem of an uncertain figure for M, comparisons between rebreathing data and more direct measurements have shown a good correspondence. Henderson and Apthorp (1960) used a breathing method with a preliminary "washout" of nitrogen from the blood and residual gas volume of the lungs. Given this careful approach, all but two of the bloodless estimates fell within ± 10% of the line of identity, relating them to direct blood determinations. Hackney et al. (1962) simplified the technique by omitting the initial nitrogen washout; nevertheless, they still found a coefficient of variation of only 4.4% for the indirect method (5.0% if the oxygen content of the rebreathing bag was assumed rather than measured).

A second and even simpler bloodless method is to measure the carbon monoxide content of alveolar gas after breath holding (R. H. Jones et al., 1958; Bergman et al., 1975; Coburn et al., 1965; Dahms and Horvath, 1974; Ringold et al., 1962). R. H. Jones et al. (1958) recommended a twenty-second period of breath holding following a maximal inspiration from the functional residual volume. At high blood carboxyhaemoglobin levels, the value predicted from the alveolar gas analysis came within 10% of the directly measured figure in 90% of samples. The results of Jones et al. (1958) suggest a disproportionately large increment of alveolar carbon monoxide pressures at high carboxyhaemoglobin readings; this has not been confirmed by other investigators (Ringold et al., 1962; McIlvaine et al., 1969). The technique apparently becomes inaccurate if there is a high atmospheric burden of carbon monoxide and the subject has not yet equilibrated with his environment (Peterson, 1970). Nevertheless, the slope of the $\overline{COHb}$%/alveolar CO line is approximately 0.155% per ppm, as predicted from the Coburn equation for carbon monoxide uptake (range 0.14 to 0.20; Coburn et al., 1965; Forbes et al., 1945; Sjöstrand, 1948a, b; Ringold et al., 1962). Several alternative linear and quadratic equations for the prediction of carboxyhaemoglobin are now available. All are based upon equilibration studies carried out at low ambient carbon monoxide pressures

$$\overline{COHb}\% = 1.8[CO,Pa] \qquad \text{McIlvaine et al. (1969)}$$
$$= 1.4[CO,Pa] + 1.09 \qquad \text{Goldsmith (1965)}$$
$$= 1.6[CO,Pa] \qquad \text{Goldsmith and Landau (1968)}$$
$$[CO,Pa] = 0.014[\overline{COHb}\%)^2 + 0.31[\overline{COHb}\%] + 4.2 \qquad \text{Peterson (1970)}$$

The third expedient is to examine mixed-expired or end-expired gas (Carlsten et al., 1954; Feldsten, 1965; Peterson, 1970; Rawbone et al., 1976; Sjöstrand, 1948; Vogt et al., 1977; Rees et al., 1980; N. J. Smith, 1977). Not all authors have found this last approach to be satisfactory. N. J. Smith (1977) stressed that the end-expired air technique became unreliable with recent smoking or exposure to a transient peak of carbon monoxide pollution. He recommended that subjects breathe a low carbon monoxide air supply (CO partial pressure < 1 Pa) for at least two minutes prior to the collection of alveolar gas samples.

Chapter 3

BIOLOGY OF CARBON MONOXIDE

In this chapter we shall examine the biochemical reactions between carbon monoxide and the body pigments and will discuss the endogenous production of carbon monoxide from haemoglobin. The dynamics of the uptake, distribution, and elimination of carbon monoxide will be reviewed, and specific comment will be made upon the exchange of carbon monoxide between the mother and the foetus.

BIOCHEMICAL REACTIONS WITH BODY PIGMENTS

General Considerations

The toxicity of carbon monoxide is due to its reaction with body pigments. The prime target is the red cell pigment haemoglobin, but as exposure continues the muscle pigment myoglobin is also affected, and if the concentration of carbon monoxide is extremely high there may also be some poisoning of cytochrome P-450 within the mitochondria (Schleyer et al., 1978).

The common feature of these three pigments is the presence of iron. Carbon monoxide has a marked tendency to react with iron and other metal atoms such as chromium, molybdenum, tungsten, and nickel. In all cases, molecules of carbon monoxide share two electrons, with the object of bringing the electron shell of the metal to a complement appropriate for the next higher stable inert gas. Thus pure iron forms a pentacarbonyl $Fe\,[CO]_5$. The reaction is helped forward by high pressure, and this can cause changes of composition when supposely standard gas mixtures are stored in cylinders for a long period (Brief et al., 1967). The carbon monoxide may also penetrate heated flue-pipes by formation of a pentacarbonyl intermediate.

Haemoglobin

Haemoglobin contains four heme molecules. Each is a ferrous iron complex of protoporphyrin IX combined with a polypeptide or globin (Shephard, 1983).

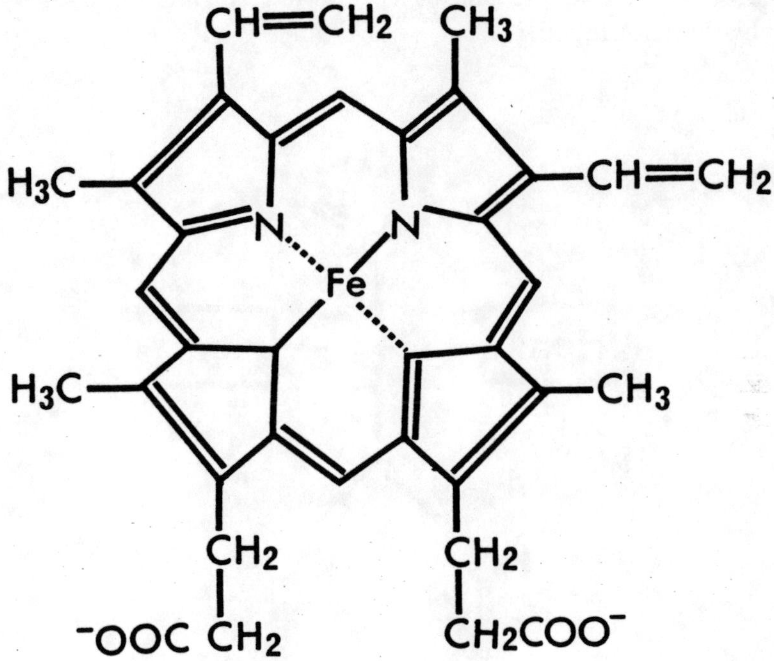

Under normal conditions, the fifth bonding position of the iron atom combines with the histidine terminal of the globin molecule, while the sixth coordination position is available to form a reversible bond with oxygen (Antweiler, 1974; Case and Karplus, 1978). Note that the oxygen bond is set at an angle to the histidine bond, this pattern suiting the steric restraints imposed by another nearby histidine terminal.

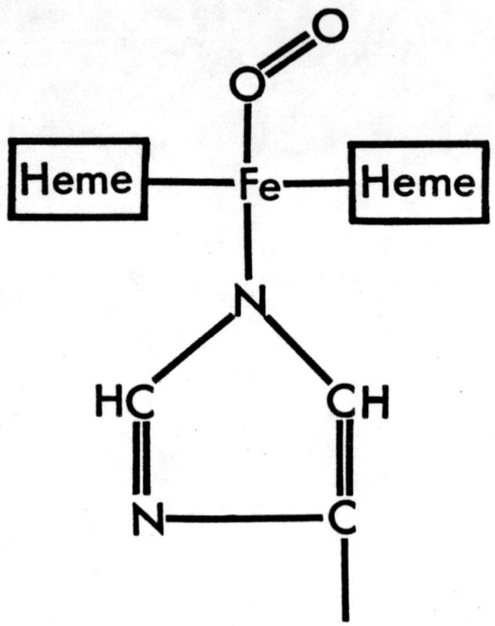

When heme is free in solution, the affinity of the iron atom for carbon monoxide is about 25,000 times that for oxygen, the carbon monoxide binding in the same plane as the iron-histidine bond:

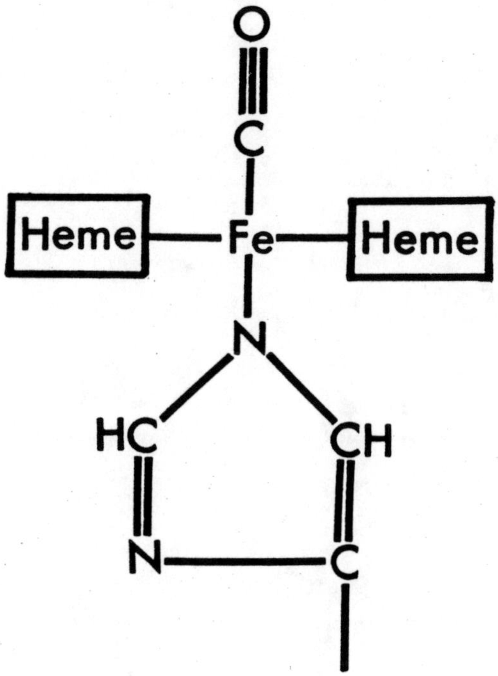

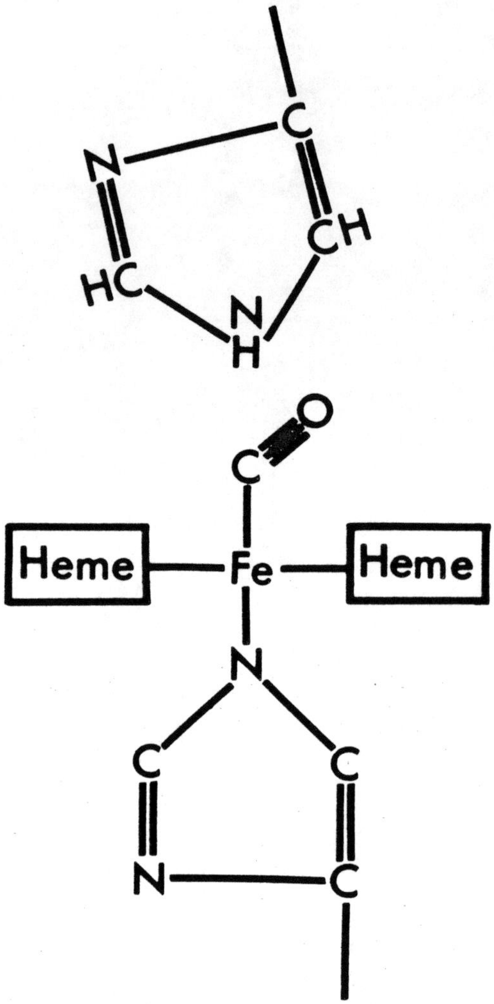

However, when the heme molecule is bound to haemoglobin, the second histidine terminal forces an angulation of the —C≡O similar to that seen for oxygen. The net effect is to maximize the affinity of the iron for oxygen and to reduce its carbon monoxide affinity to the M value of 220 to 250 seen *in vivo* (Douglas et al., 1912; Joels and Pugh, 1958; Antonini, 1967; Perutz, 1970; Roughton, 1964, 1970; Rodkey et al., 1974; see Chapter 2). In a teleological sense, this can be viewed as an important protection of the body against endogenous production of carbon monoxide. Even with this protection, the endogenous output of carbon monoxide is sufficient to block about 1% of the binding sites of both haemoglobin and myoglobin.

The combination of carbon monoxide with haemoglobin was first discovered by the French physiologist Claude Bernard; he ascribed death to anoxaemia, due to the displacement of oxygen from the haemoglobin molecule. Haldane (1895a) confirmed the basic hypothesis, showing that a cockroach (which has no haemoglobin) could live happily in 80% carbon monoxide and 20% oxygen for a fortnight, while a mouse also survived exposure to 1 atmosphere of carbon monoxide if oxygen was supplied at sufficient pressure (2 atmospheres) to allow tissue demand to be met by its transport in physical solution. Nevertheless, anoxaemia did not seem the entire explanation of CO poisoning. As Haldane and Priestley pointed out in 1935 "If the action of carbon monoxide were simply to diminish the oxygen carrying capacity of the blood. . . . the symptoms of CO poisoning would be very difficult to understand. . . . a person whose blood is half saturated with CO is practically helpless . . . but a person whose haemoglobin percentage is simply diminished to a half by anaemia may be going about his work as usual."

The primary explanation of this discrepancy is a CO-induced change in the oxygen dissociation curve of haemoglobin (Haldane and Priestley, 1935; Roughton and Darling, 1944; Mulhausen et al., 1968; Collier, 1976; Ledwith, 1978) "such . . . that the haemoglobin holds on more tightly to the oxygen" (see Figure 3); "with 50 percent saturation of the blood with CO, the oxygen pressure must fall to less than half the usual value . . . in order to dissociate half

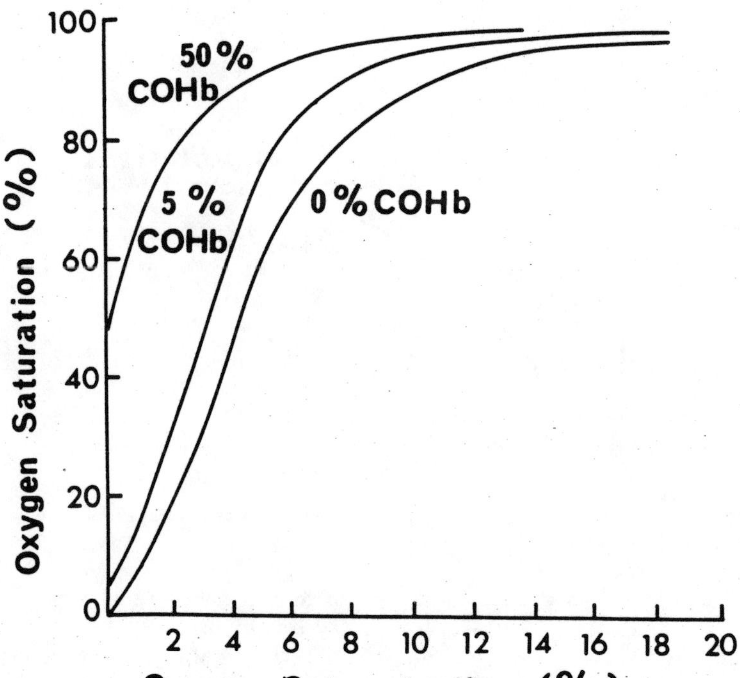

Figure 3. Influence of carboxyhaemoglobin formation upon the shape of the oxyhaemoglobin dissociation curve. Based upon data of Haldane and Priestley, 1935.

the oxygen present . . ." (Haldane and Priestley, 1935). The lowering of the mixed venous oxygen pressure relative to a 50% anaemia is illustrated in Figure 4. The more modest levels of carboxyhaemoglobin commonly encountered in cigarette smokers and city dwellers still induce a substantial change in the oxygen pressure needed for 50% saturation of haemoglobin (Le Groumellec et al., 1978). Thus Mulhausen et al. (1968) noted a drop of the "P_{50}" reading from 3.56 to 3.09 kPa. Patients with sickle-cell anaemia seem particularly vulnerable to the CO-induced change in the shape of the oxygen dissociation curve (Cameron et al., 1972). The transition from a sigmoid to a hyperbolic curve reflects an elimination of the first site of heme oxygenation rather than any more fundamental change in the structure of the haemoglobin molecule. Normally, the combination of an atom of oxygen with the first heme unit alters the shape of the haemoglobin molecule, facilitating oxygenation of the remaining three heme units; presumably, if the first combination is between heme and carbon monoxide, there is less facilitation of subsequent heme/

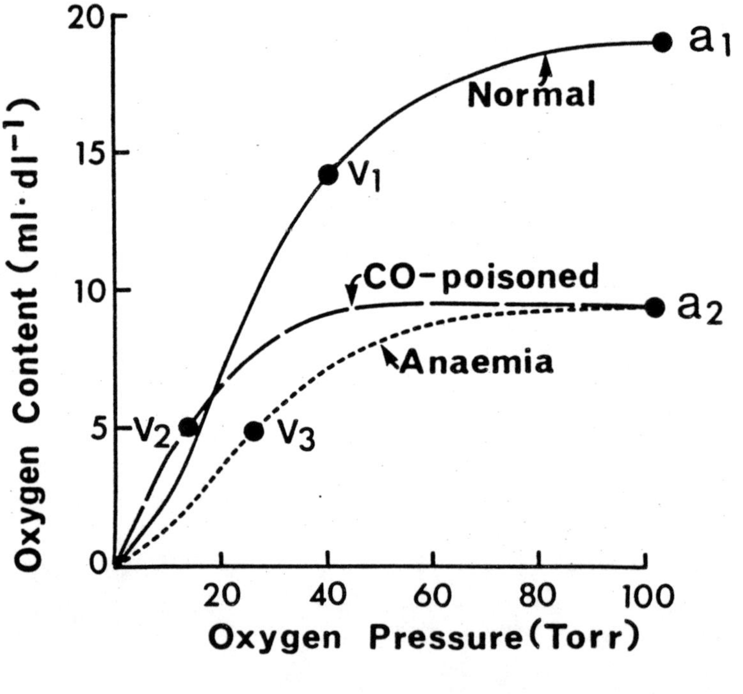

Figure 4. A comparison of mixed venous oxygen content and venous oxygen pressures for normal resting conditions (V_1—arteriovenous oxygen difference 5 ml·dl^{-1}), 50% carboxyhaemoglobin (V_2), and 50% anaemia (V_3). Based on the curves of Roughton and Darling, 1944. In practice, the situation of the poisoned subject may be even worse, due to some widening of the alveolar-arterial oxygen pressure gradient.

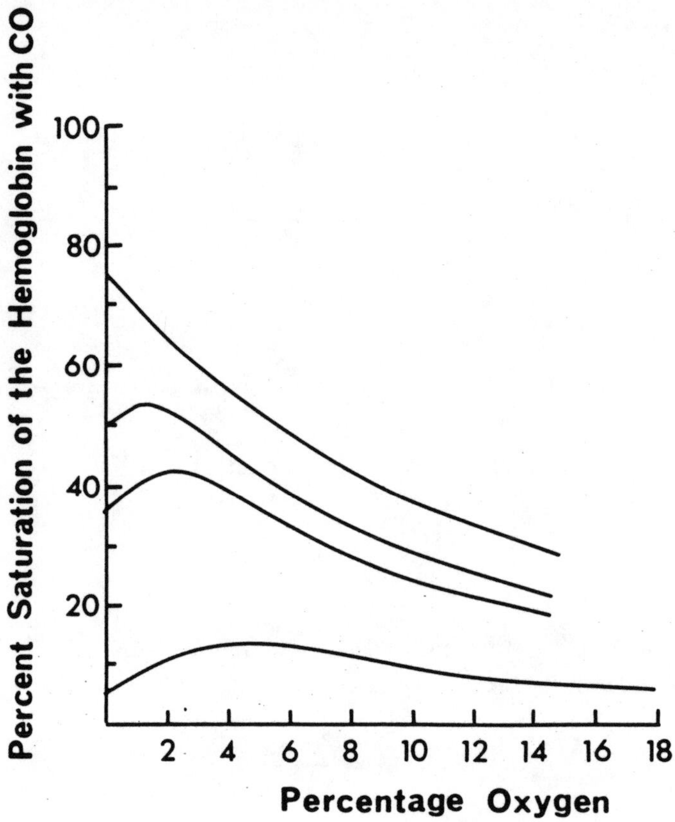

Figure 5. Influence of partial pressure of oxygen upon the reaction between carbon monoxide and haemoglobin. Based on the data of Haldane and Priestley, 1935.

oxygen interactions (Hlastala et al., 1976; Okada et al., 1976; Geibel et al., 1978; Sawicki and Gibson, 1978).

Because of competition between oxygen and carbon monoxide for the heme binding sites, an increase in the partial pressure of oxygen normally reduces the percentage of carboxyhaemoglobin. However, if the initial sum of oxygen and carbon monoxide pressures is insufficient for full saturation of haemoglobin, an increase of oxygen pressure increases the binding of carbon monoxide (Joels and Pugh, 1958; see Figure 5); this probably reflects exposure of remaining heme units by the process of oxygen binding (Stryer, 1981).

Both the reactions of oxygen with haemoglobin (Adair, 1925) and the analogous reactions for carbon monoxide (Holland, 1969 a, b; Kernohan, 1961) can be represented by a sequence of reactions with the four heme units:

$$Hb_4 + O_2 \underset{K_1}{\overset{K_1'}{\rightleftharpoons}} Hb_4O_2$$

$$Hb_4O_2 + O_2 \underset{K_2}{\overset{K_2'}{\rightleftharpoons}} Hb_4O_4$$

$$Hb_4O_4 + O_2 \underset{K_3}{\overset{K_3'}{\rightleftharpoons}} Hb_4O_6$$

$$Hb_4O_6 + O_2 \underset{K_4}{\overset{K_4'}{\rightleftharpoons}} Hb_4O_8$$

The equilibrium constant M (Douglas et al., 1912) in the equation

$$\frac{[HbCO]}{[HbO_2]} = M\frac{Pco}{Po_2}$$

depends upon the relative magnitude of the various reaction constants, particularly those for reaction with the fourth heme unit

$$M = \frac{l'_4 \cdot K_4}{K_4' \cdot l_4}$$

where l'_4 and l_4 are forward and reverse reaction constants for the binding of carbon monoxide. The association velocity constants for carbon monoxide are only one-fourth to one-sixth of those for oxygen, and the high value of M is attributable to a slow dissociation constant (1/800th of that for oxygen; Forster, 1964).

Some authors have found a variation of M with pH (Allen and Root, 1955a; Joels and Pugh, 1958), with values rising from 180 at a pH of 7.15 to a maximum of 230 at a pH of 7.35 and falling back to 140 at a pH of 7.70. Others find no change with pH (Douglas et al., 1912; Rodkey et al., 1969), oxygen pressure (Douglas et al., 1912), or levels of red cell 2:3 diphosphoglycerate (2:3 DPG: Benesch and Benesch, 1967; Okada et al., 1976). However, the affinity ratio is altered by temperature (Sirs, 1976) and differs over various segments of the oxygen dissociation curve (Roughton, 1970). Hartridge and Roughton also noticed that when a dilute solution of carboxyhaemoglobin was exposed to strong light, the absorption spectrum became that of oxyhaemoglobin. This was the first evidence of the photolysis effect discussed above (Gibson and Ainsworth, 1957; Roughton et al., 1963; Gibson, 1959; Bernard et al., 1975; Noe et al., 1978; Pandolfelli et al., 1978); the binding of oxygen to haemoglobin has only one-thirtieth the photosensitivity of the carbon monoxide-haemoglobin bond.

The normal response to hypoxia is an alteration in the oxygen-binding properties of haemoglobin, an increase of 2:3 DPG shifting the dissociation curve to the right (Stryer, 1981; H. M. Thomas et al., 1974). It is still debated whether such an adaptation occurs in chronic carbon monoxide poisoning; Mulhausen et al. (1968) have reported no change in the oxygen-binding properties of haemoglobin under such circumstances, Åstrup (1970) a small decrease, and Dinman et al. (1970) a small increase of 2:3 DPG.

Myoglobin

Myoglobin has a similar general structure to haemoglobin, although there is only one heme unit and the globin differs slightly from that found in the red cell pigment. It is thus not surprising that both oxygen and carbon monoxide also react with the muscle pigment. Diffusion of carbon monoxide from the circulation to muscle has indeed been recognised for a long time as a source of error when using CO to estimate total blood volume. At equilibrium, some 10% of the inhaled dose of carbon monoxide is bound to extravascular pigment.

The affinity of carbon monoxide for myoglobin and other intracellular compounds is usually expressed as the Warburg partition coefficient K. This describes the ratio of carbon monoxide to oxygen pressures required to produce a 50% saturation of the pigment with carbon monoxide. It is analogous to the reciprocal of M. Reported values of K for myoglobin range from 0.025 (Coburn, 1971; Coburn et al., 1973; Luomanmaki, 1966; Rossi-Fanelli and Antonini, 1958) to 0.05 (Millikan, 1939). During an acute exposure, the rise of carboxymyoglobin level nevertheless lags behind that of carboxyhaemoglobin (Godin and Shephard, 1972). Under steady-state conditions, blood and muscle pigment saturations tend to equalize (since $K \approx 1/M$), but hypoxia can increase the myoglobin/haemoglobin ratio from 1 to around 3, both for cardiac and skeletal muscle (Coburn et al., 1973; Coburn and Mayers, 1971; see Figure 6).

If the oxygen pressure falls in the tissue capillaries (as during maximum exercise), carbon monoxide is displaced from blood to muscle (Coburn et al., 1973; Luomanmaki, 1966; Clark and Coburn, 1975). The oxygen pressure is usually a little lower in cardiac tissue (0.5 to 0.8 Pa) than in skeletal muscle, and a 10% blood carboxyhaemoglobin reading can lead to a 30% carboxymyoglobin level in heart muscle.

The normal functions of myoglobin are to provide a temporary store of oxygen for brief bursts of interval work and to facilitate the diffusion of oxygen within the muscle cells (Wilks, 1963). It is likely that both functions are disturbed in proportional fashion as carboxymyoglobin is formed.

Cytochrome Pigments

The possibility that carbon monoxide could disturb intracellular electron

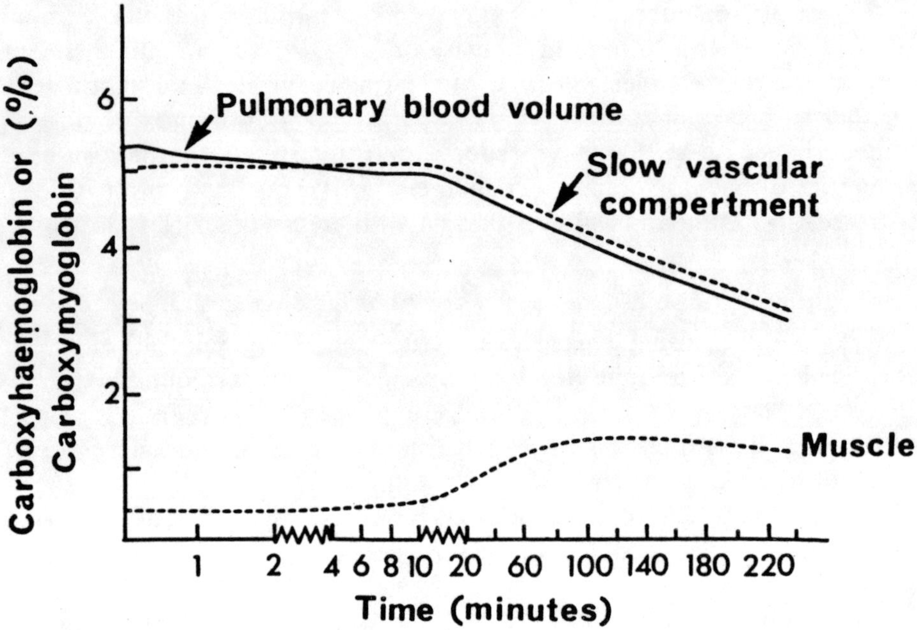

Figure 6. Distribution of carbon monoxide in the period following the rebreathing of a high concentration of carbon monoxide for two minutes. Based on data obtained by Godin and Shephard, 1972, using an analogue.

transport was first noted by Warburg (1926). He observed that the "Atmungsferment" responsible for the oxygen consumption of yeast was depressed when the partial pressure of carbon monoxide was five times that for oxygen. J. B. S. Haldane (1927) showed a similar effect upon germinating cress and in moths. Nevertheless, it has been argued that at usually encountered CO pressures, the combination of carbon monoxide with the various cytochrome pigments is insufficient to be of practical significance (Root, 1965). While this view is probably correct for equilibrium conditions, if a bolus of carbon monoxide is administered, the finite rate of reaction with haemoglobin and myoglobin allows a temporary attack upon other tissue pigments (Smialek, 1974; White, 1977; D. Y. Cooper et al., 1977; Goldbaum, 1977). Some of the animal experiments where such an effect has been described have used carbon monoxide concentrations of up to 10 kPa.

A critical and somewhat uncertain variable is the intracellular oxygen pressure. Studies of carbon monoxide binding to myoglobin and the use of polarographic microelectrodes have indicated intracellular oxygen pressures in the range 0.5 to 0.8 kPa (Coburn et al., 1973; Kessler et al., 1973; Whalen, 1971), but at some points within the cell pressures may be as low as 1 Pa

(Shephard, 1977), particularly in the presence of carboxyhaemoglobin. In the rabbit, peritoneal carbon monoxide pressures reach to 42 to 69% of values found in alveolar gas (Gothert et al., 1970). At a blood carboxyhaemoglobin level of 5% and a mean tissue capillary oxygen pressure of 6 kPa, the carbon monoxide pressure is probably 2 to 3 Pa.

The Warburg coefficient for cytochrome a_3 varies from 2 to 30 (Ball et al., 1951; Coburn, 1977; Wohlrab et al., 1971; Neil and Joels, 1963). Early workers estimated that the CO pressure needed to produce 50% saturation of "cytochrome oxidase" (a mixture of cytochrome a and cytochrome a_3) was in the range of 21 to 29 Pa (Gibson and Greenwood, 1964; Wald and Smith, 1974). Interpretation of data is complicated by the fact that measurements were usually made at 25° C rather than 37° C. Since the tissue P_{CO} with a 5% blood carboxyhaemoglobin level is 2 to 3 Pa and K for cytochrome $a_3 = 2$, the mitochondrial oxygen pressure would need to drop to 1.3 Pa before 50% binding occurred (Longo, 1977). In essence, carbon monoxide binds only with reduced cytochrome a_3. Since there is also an excess of cytochrome a_3 in the mitochondria, a combination of a high carboxyhaemoglobin reading ($>$ 10%) and tissue hypoxia (P_{O_2} $<$1.3 Pa) are necessary before binding assumes physiological significance. Forster (1969a) suggests that oxygen consumption remains unaffected even if 50% of the cytochrome a_3 is inhibited. Nevertheless, an increase of mitochondrial cytochrome a_3 could conceivably play a role in acclimatization to very high partial pressures of carbon monoxide.

The Warburg coefficient for cytochrome P-450 varies from 0.2 to 5.0 (Estabrook et al., 1970). Similar calculations to those made for cytochrome a_3 suggest that this component of the electron transport chain becomes sensitive to carbon monoxide poisoning when metabolism is rapid (as in vigorous aerobic exercise), but it is unaffected by CO under normal resting conditions (Estabrook et al., 1970). Examples of impaired cellular function include a decreased hepatic metabolism of 3-OH-benzo alphapyrene in rats exposed to 6 Pa of carbon monoxide (Rondia, 1970) and a decreased catabolism of haemoglobin-haptoglobin in dogs with a carboxyhaemoglobin level of 10 to 12% (Coburn and Kane, 1968).

It remains unclear how far tissues can compensate for poisoning of the mitochondrial enzymes by an increase of perfusion and thus of tissue oxygen pressure. This type of adaptation seems likely under resting conditions but is much less feasible during vigorous exercise. One argument against any major compensatory increase of perfusion is a *fall* of tissue oxygen pressure during carbon monoxide poisoning (Weiss and Cohen, 1974; Zorn, 1972).

If the oxygen pressure were to drop low enough, carbon monoxide could theoretically interact with the reduced form of other iron containing enzymes such as catalase and peroxidase. However, such reactions are exceedingly unlikely in practice, since these compounds are not normally found in a reduced form within the body.

ENDOGENOUS PRODUCTION OF CARBON MONOXIDE

The urban nonsmoker commonly has a blood carboxyhaemoglobin level of ~ 1% (Coburn et al., 1965; Rode and Shephard, 1971). However, only about half of this burden is attributable to atmospheric pollution, the residue reflecting an endogenous production of carbon monoxide (Sjöstrand, 1949, 1951; Bensinger et al., 1975) during the breakdown of haemoglobin and haem-containing enzymes.

About three-quarters of the endogenous production of 5-10 $\mu l \cdot kg^{-1} h^{-1}$ (Coburn et al., 1963; Luomanmaki, 1966; Delivoria-Papadopoulos et al., 1974; Lynch and Moede, 1972) is derived from the alpha methene carbon atom of the heme molecule (Coburn et al., 1967; Coburn et al., 1964; Lundwig and Blakemore, 1957; Sjöstrand, 1952 a, b; Logue et al., 1971). The process of erythrocyte breakdown (and thus carbon monoxide formation) can be increased several-fold by such situations as fasting (Mercke et al., 1975), administration of drugs such as phenobarbitol and diphenylhydantoin that induce hepatic cytochrome (Coburn, 1970), menstruation (Delivoria-Papadopoulos et al., 1970), pregnancy (Longo et al., 1967), elevation of plasma progesterone (Delivoria-Papadopoulos et al., 1974), sustained exercise (Shephard, 1982b), simulated saturation dives (Rodkey et al., 1974), and hemolytic disease (Coburn et al., 1966; Engel et al., 1971; Sjöstrand, 1949). The pregnant woman is exposed to not only her own carbon monoxide production but foetal carbon monoxide output exchanged across the placenta; the latter accounts for around 3% of the total carboxyhaemoglobin in the blood of a nonsmoking pregnant woman (Åstrup et al., 1972; Longo et al., 1967; Longo, 1970).

Some carbon monoxide is also produced during the breakdown of hepatic, heme-containing enzymes (Coburn, 1973). However, the turnover of myoglobin is relatively slow, and it probably contributes little to the endogenous burden of carbon monoxide (Forster, 1969b).

A few industrial chemicals are broken down to carbon monoxide within the body. The most important is methylene chloride (CH_2Cl_2), used in household aerosols and paint strippers (Ratney et al., 1974). This substance is absorbed and oxidized to carbon monoxide. Inhalation of the vapour at the concentrations likely when removing paint (500 to 1000 ppm) can produce blood carboxyhaemoglobin readings as high as 10% (R. D. Stewart et al., 1972; Miller and Engel, 1973; Kubic and Anders, 1974; Ratney and Wegman, 1974; Carlsson and Huttengren, 1975; Stewart and Hake, 1976; Barrowcliff, 1978; Benzon et al., 1978; Boudene et al., 1978; Laham et al., 1978; Stevenson et al., 1978; Barrowcliff and Knell, 1979). Details of the biological reactions involved are uncertain, although it is agreed that the process is quite slow, with carboxyhaemoglobin readings continuing to increase for as much as one hour after exposure has ceased. Methylene chloride itself has no irritant properties to warn the painter of the danger he is facing, and the only remedy (other than adequate ventilation) is to add an irritant to the solvent that makes a toxic concentration of the mixture unacceptable to personal comfort.

UPTAKE, DISTRIBUTION, AND ELIMINATION OF CARBON MONOXIDE

The fact that man is exposed to constantly varying concentrations of carbon monoxide has led to a substantial interest in the development of equations to predict the uptake, distribution, and rate of elimination of carbon monoxide. Of particular concern has been the blood carboxyhaemoglobin reading resulting from several minutes' to several hours' exposure to heavily contaminated air, whether in the urban or the industrial environment. Unfortunately, the constants needed for such predictions have generally been derived from relatively small groups of subjects, often using poorly controlled techniques. End results thus differ substantially between investigators. Moreover, most studies have been conducted under resting conditions; there are occasional reports covering the levels of physical activity encountered in industry, but there is no information on vigorous exercise (when CO uptake is greatly speeded by a large respiratory minute volume and cardiac output).

Uptake and Distribution

Almost all of the carbon monoxide entering the body is rapidly bound to heme-containing pigments (Coburn, 1967). At least 80% is bound to haemoglobin, 10 to 15% reacts with myoglobin, and 5% can react with other heme-containing compounds (particularly in the liver). As discussed above, the carbon monoxide is most likely to penetrate to tissue pigments when the peripheral oxygen pressure is low (Coburn and Mayers, 1971; Luomanmäki and Coburn, 1969).

The rate of build up of carboxyhaemoglobin depends on the inspired carbon monoxide pressure and the volume of air breathed per minute relative to body mass (and thus blood volume; Haldane, 1895a; Henderson and Haggard, 1921). The combined effects of airway dead space, slow absorption at the alveolar membrane, and back-pressure at the surface of the red cell are such that only about half of the respired dose of carbon monoxide is retained under resting conditions (Haldane, 1895a; Forbes et al., 1945; Bates, 1952). During light physical activity, the initial fractional uptake of carbon monoxide remains around 50%, but during heavier exercise (a power output of 100 Watts, corresponding to a heart rate of 135 beats$\cdot$min^{-1}), retention drops to about 40% (Forbes, 1970; Lilienthal and Pine, 1946). A rising back-pressure of carbon monoxide in the pulmonary capillaries causes a slowing of uptake, which becomes appreciable when subjects have moved more than a third of the way towards the equilibrium level for a given environment (Forbes et al., 1945; Haldane, 1895a). In a healthy young adult performing light physical activity,

Table VIII
CARBOXYHAEMOGLOBIN LEVELS OF A HEALTHY RESTING NONSMOKER EXPOSED TO KNOWN PARTIAL PRESSURES OF CARBON MONOXIDE FOR SPECIFIED PERIOD*

Partial pressure of respired carbon monoxide	Duration of Exposure and Blood Carboxyhaemoglobin (%)				
	10 minutes	30 minutes	1 hour	8 hours	Equilibrium
Endogenous CO	0.4	0.4	0.4	0.4	0.4
0.5 Pa			0.5	0.8	0.9
1 Pa		0.5	0.6	1.7	1.8
1.5 Pa		0.6	0.7	2.4	2.7
2 Pa		0.7	0.9	3.0	3.5
3 Pa	0.5	0.8	1.3	4.5	5.0
4 Pa	0.6	1.0	1.5	5.9	6.1
5 Pa	0.7	1.3	1.9	7.0	7.8
7.5 Pa	0.8	1.7	2.6	9.6	11.0
10 Pa	1.1	2.2	3.0	12.5	14.0
20 Pa	1.9	4.0	6.0	20.0	24.6

*Values estimated from data of Forbes et al., 1947, and Peterson and Stewart, 1970.

the half-time for carbon monoxide uptake is about two hours (Brody and Coburn, 1970). Under resting conditions, equilibrium is reached more slowly, particularly in older individuals (where the pulmonary diffusing capacity is decreased by aging, emphysema, and/or chronic obstructive lung disease, (Bates, 1952; Ramsey, 1972; see Table VIII).

Elimination

Carbon monoxide is eliminated almost entirely via the lungs. At most, 1% of the gas is oxidized within the body (Fenn and Cobb, 1932; Luomanmäki and Coburn, 1969; Luomanmäki, 1966; Roughton and Root, 1945b), although some redistribution may also occur to other body compartments (Roughton and Root, 1945b).

Most investigators have constrained their data to a single exponential curve. Under resting conditions, the resultant half-time of carbon monoxide elimination is generally accepted as about three to four hours for a subject breathing air and about one hour breathing oxygen at normal atmospheric pressures (Bartlett, 1968; Pace et al., 1950; Roughton and Root, 1945); somewhat faster rates are observed at very high carboxyhaemoglobin concentrations (Haldane, 1895a; Henderson et al., 1921) and slower rates at very low concentrations (Peterson and Stewart, 1970). Some authors have observed rather more rapid half-times (Norman and Ledingham, 1967; Roughton and Root, 1945b), particularly where small doses have been involved.

Rose et al. (1972) pointed out two problems with most reports. Half-times were calculated without reference to the normal urban background of ~1% carboxyhaemoglobin. This must inevitably lead to departure from a simple exponential function unless data is expressed as a *difference* from background. Similar rates of carbon monoxide elimination have also been assumed for men and women. However, Rode et al. (1972) found half-times of 3.7 hours for men and 2.5 hours for women; several other authors have confirmed faster carbon monoxide transport in women (Pace et al., 1950; Goldsmith et al., 1963; Peterson and Stewart, 1975). Other potential variables include the age and health of the subjects, altitude (Myhre, 1974), possible redistribution of carbon monoxide into poorly perfused vascular compartments such as the spleen and bone marrow, and the need to allow for reactions with extravascular pigments (Godin and Shephard, 1972; Killick and Marchant, 1959; Roughton and Root, 1945; Tobias et al., 1945). Failure to allow for these factors probably accounts for the more variable and concentration-dependent half-times reported by some investigators (Coburn et al., 1965; Mercke et al., 1975; Myhre, 1974).

Roughton and Root (1945b) commented that only 60 to 70% of the gas leaving the blood in the first hour after exposure was expired, implying a surprisingly large redistribution of CO. More recently, Godin and Shephard (1972), Wagner et al. (1975), and Myhre (1974) have all noted a biphasic carbon monoxide elimination curve. Presumably, much depends upon the CO pressure to which the individual is exposed and upon the sophistication of data analysis; with adequate methodology, a multicomponent curve might be described. Godin and Shephard (1972) devised an analogue based upon the probable regional distribution of blood flow and the volumes of the various tissue compartments. In this model, carboxymyoglobin concentrations rose steadily during six minutes of exposure to a high concentration of carbon monoxide, and under such conditions there was an "overshoot" of carboxyhaemoglobin concentrations in the pulmonary capillary blood and other components of the rapidly circulating blood volume. During the first five minutes of the recovery period, equilibrium between the various blood compartments was completed, but carbon monoxide continued to penetrate into the myoglobin for up to one hour (see Figure 6).

While most authors continue to advocate exponential clearance models, Wagner et al. (1975) found that their data for the period fifteen to ninety minutes after exposure to a substantial dose of carbon monoxide were best satisfied by a linear equation. During this phase of carbon monoxide elimination, clearance depended upon the affinity constants for haemoglobin and myoglobin rather than the respiratory minute volume or the cardiac output.

Henderson and Haggard (1921) argued that breathing was inadequate after CO poisoning due to a combination of hyperventilation and CO_2 loss during exposure, with a subsequent temporary "exhaustion" of the respiratory centre.

The exhaustion is more accurately a depression of the respiratory control mechanism by oxygen lack (Donald and Paton, 1955). Nevertheless, there may also be a respiratory alkalosis secondary to metabolic acidosis (Norman et al., 1966), so that clearance of carbon monoxide apparently proceeds most rapidly if both circulatory and ventilatory systems are stimulated by either exercise (Song and Eagan, 1978) or breathing a mixture of 5% carbon dioxide in oxygen (Killick, 1936; Marriott, 1955, 1958; Medical Research Council, 1958; Killick and Marchant, 1959; Apthorp et al., 1958; Malorney et al., 1962; Antweiler, 1974). No additional advantage is gained from the use of a 7% CO_2 mixture (Norman and Ledingham, 1967), and indeed oxygen alone is equally effective in the resuscitation of CO-poisoned subjects if ventilation is increased by mechanical means (Apthorp et al., 1958).

Modern concepts of hyperbaric therapy can be traced back to the work of Haldane (1895a), who demonstrated that a mouse could remain quite normal while breathing a mixture that contained two parts of oxygen and one part of carbon monoxide, provided that the ambient pressure was increased to two atmospheres. Benefit was derived in part from the oxygen that dissolved in physical solution in the blood and in part from a dissociation of the carboxyhaemoglobin induced by the increased partial pressure of oxygen (G. Smith, 1963). The half-time of carbon monoxide elimination by a human subject is reduced to approximately twenty-three minutes if the affected individual is placed in a hyperbaric chamber at an oxygen pressure of 3 atmospheres (Kindwall, 1977; Pace et al., 1950; G. K Anderson, 1978); at the same time, any tissue hypoxia is corrected by the hyperbaric conditions (Norman and Ledingham, 1967).

One interesting new technique for the treatment of severe poisoning is asanguineous hypothermic perfusion (Agostini and Ramirez, 1974).

Models of Carbon Monoxide Uptake and Elimination

Models proposed to predict the uptake and release of carbon monoxide from the body range from simple to quite complex formulae (Chovin, 1974; see Table IX).

One of the earliest equations (Forbes et al., 1945) was based upon data obtained during six-hour exposures to relatively high partial pressures of carbon monoxide (10 Pa to 20 kPa)

$$\text{Rate of CO uptake} = \text{KP}_{\text{CO}}\frac{\overline{\text{COHb}}_\infty - \overline{\text{COHb}}_t}{\overline{\text{COHb}}_\infty - \overline{\text{COHb}}_o}$$

In this expression, K was a constant proportional to the intensity of physical activity, P_{CO} was the partial pressure of carbon monoxide in inspired air, and $\overline{\text{COHb}}$ was the percentage of carboxyhaemoglobin initially (o), at any given

Table IX
PREDICTED BLOOD CARBOXYHAEMOGLOBIN READINGS DURING EXPOSURE TO A CARBON MONOXIDE PRESSURE OF 10 Pa USING MODELS OF SEVERAL AUTHORS

	Activity status and duration of exposure					
	Rest			Light exercise		
Author	15 min	60 min	480 min	15 min	60 min	480 min
Coburn et al. (1965)	1.2	3.0	12.4	2.0	5.7	15.9
Forbes et al. (1945)	1.0	2.3	14.4	1.7	5.3	38.4
Hanks and Farquhar (1969)	1.1	2.8	10.9	1.9	5.4	14.6
Pace et al. (1946)	0.8	1.9	11.5	1.6	4.7	34.1
Peterson and Stewart (1970)	1.5	3.5	13.0	—	—	—

time (t), or at equilibrium (∞), as indicated by the respective subscripts. The increase of carboxyhaemoglobin ($\Delta\ \overline{COHb}\%$) was then given by the formula

$$\Delta\ \overline{COHb}\% = K'(\%CO)_t$$

where K′ was a factor increasing nonlinearly with heart rate and respiratory minute volume from 3 to 11, and t was the duration of exposure in minutes. Assuming a normal initial carboxyhaemoglobin reading (less than 5%), the equation yielded valid results at least over the early part of equilibration (a $\Delta\ \overline{COHb}$ of 7% at 10 Pa or 30% at 100 Pa of carbon monoxide). Larsen and Burke (1969) supported the use of this type of equation, recommending K′ values of 3 for resting conditions, 8 for light to moderate work (respiratory minute volume 18 $l \cdot min^{-1}$), and 16.7 for heavy work. Lilienthal and Pine (1946) proposed an analogous formula on the basis of results collected during rest and light exercise at ambient carbon monoxide pressures ranging from 5 Pa to 275 Pa.

$$\Delta\ \overline{COHb}\% = 0.05\ \dot{V}_E\ (P_{CO})t$$

where $\dot{V}_E$ was the respiratory minute volume ($l \cdot min^{-1}$) and P_{CO} was measured in Torr. Expressing the carbon monoxide pressure as a percentage of an atmosphere, this would yield a resting K′ value of 2.28($\dot{V}_E = 6 l \cdot min^{-1}$) and a K′ of 11.4 for moderate activity ($\dot{V}_E = 30 l \cdot min^{-1}$).

The rate of carboxyhaemoglobin formation is necessarily inversely proportional to blood volume; Pace et al. (1946) thus introduced blood volume into their equation.

$$\Delta\ \overline{COHb}\% = \frac{100\ \dot{V}_E\ (\%CO)\ t}{46.5\ (\text{blood volume, litres})}$$

Their formula was suggested as valid for carbon monoxide partial pressures of 10 to 200 Pa, from the onset of exposure until the blood carboxyhaemoglobin readings reached one-third of equilibrium values. The standard deviation of the estimate was said to be 2.3% COHb. Results were identical with those obtained by the equation of Forbes et al. (1945) if effective blood volumes of 4.3, 4.1, 4.9, and 5.9 litres were assumed for the four work rates tested.

Goldsmith et al. (1963) reasoned that the nature of uptake and release called for exponential rather than linear equations. The functions that they suggested were

$$\text{Uptake } (\Delta\, \overline{COHb}_t) = (\overline{COHb}_\infty - \overline{COHb}_0)(1-e^{-0.004t})$$
$$\text{Release } (\Delta\, \overline{COHb}_t) = (\overline{COHb}_0 - \overline{COHb}_\infty)(1-e^{-0.0033t})$$

The uptake constant of 0.004 was derived from a graphic analysis of the data of Forbes et al. (1945), while the excretion constant of 0.0033 was selected to yield a half-time of 3.5 hours for a return to normal resting carboxyhaemoglobin readings. Although the equations of Goldsmith et al. (1963) are widely used in the United States, they have certain inherent weaknesses. In particular, no allowance is made for varying levels of physical activity, and the base data refer to high partial pressures of carbon monoxide. The description of long-term changes in COHb% is relatively accurate, but predictions are much less precise for short-term transients. Hanks and Farquhar (1969, cited in WHO, 1979) developed an exponential formula analogous to the Forbes approach.

$$\Delta\, \overline{COHb}\% = 1{,}470\ (CO\%)\,(1 - e^{-0.00289t})$$

Putting t = 480 minutes (as in Forbes experiments), K′ would amount to 2.3, although over a shorter interval (t = 60 minutes), it would rise to 3.9. Again no allowance is possible for varying intensities of physical activity. Chovin (1974, cited in WHO, 1979) proposed rectifying this deficiency with the formula

$$\Delta\, \overline{COHb}\% = 1{,}470\ (CO\%)\ (1 - e^{-Kt/147})$$

where K = 0.025 at rest, increasing to K = 0.065 in heavy physical work.

Peterson and Stewart (1970) proposed an empirical logarithmic format of the type

$$\log(\Delta\, \overline{COHb}\%) = 0.857 \log\,[C_{I.CO}\text{ ppm}] + 0.629 \log\,(t,\text{min}) - 2.295$$

If arranged in the format of the earlier equations, it would read

$$\Delta\, \overline{COHb}\% = 0.0051\,[C_{I.CO}\text{ ppm}]^{0.858}[t]^{0.629}$$

One attraction of their formula is that the subjects (eighteen healthy graduate students) were exposed to much more realistic carbon monoxide pressures (0.1 to 100 Pa) for periods of thirty minutes to twenty-four hours. Disadvantages were a limitation of experiments to sedentary conditions and constant carbon monoxide pressures rather than the sharp ambient peaks of pollution encountered in industry and in urban driving. Finding a rather long estimated half-time for carbon monoxide elimination (128 to 409 minutes, average 320 minutes), Peterson and Stewart modified their equation to allow for this factor.

$$\Delta\,CO\overline{Hb}\% = \frac{CO^{0.858}\,(ppm)\cdot t^{0.63}\,(min)}{197}\,10^{-0.00094t}$$

The standard error of the revised equation was 1.18%$CO\overline{Hb}$. When partial pressures of carbon monoxide were high, an alternative formula was suggested (Peterson and Stewart, 1972):

$$\text{Log } (\Delta\,CO\overline{Hb}\% \text{ per litre of air breathed}) = 1.036 \log (C_{I,CO}\,ppm) - 4.4793$$

This formula would imply a $\Delta\,CO\overline{Hb}$ of 14% over 380 minutes, given a CO pressure of 10 Pa and a respiratory minute volume of $7.1\,l \cdot min^{-1}$.

Coburn et al. (1965) developed a more sophisticated equation that incorporated many pertinent physiological variables, including the pulmonary capillary oxygen pressure ($P\bar{c},O_2$), the haemoglobin affinity ratio (M), the oxygen saturation ($O_2Hb\%$), the pulmonary diffusing capacity ($\dot{D}_L$), the alveolar ventilation ($\dot{V}_A$), the endogenous production of carbon monoxide ($\dot{V}_{CO}$), the blood volume (V_b), and the total inspired gas pressure (P_I). The basic differential equation for body CO uptake ($\delta CO/\delta t$) was

$$\frac{\delta[CO]}{\delta t} = \dot{V}_{CO} - \frac{[CO\overline{Hb}]}{[O_2Hb]}\cdot\frac{P\bar{c},O_2}{M}\left[\frac{1}{\frac{1}{\dot{D}_L}+\frac{P_I-47}{\dot{V}_A}}\right] + \left[\frac{P_{I,CO}}{\frac{1}{\dot{D}_L}+\frac{P_I-47}{\dot{V}_A}}\right]$$

$$\text{Putting } A = \frac{P\bar{c},O_2}{M[O_2Hb]} \text{ and } B = \frac{1}{\dot{D}_L} + \frac{P_I - 47}{\dot{V}_A}$$

it can be shown that

$$\frac{\delta[CO]}{\delta t} = [\dot{V}_{CO}]\cdot B - [CO\overline{Hb}]\,A + P_{I,CO}$$

and at time t

$$\frac{A[\overline{COHb}]t - [\dot{V}_{CO}]B - P_{I,CO}}{A[\overline{COHb}]_o - [\dot{V}_{CO}]B - P_{I,CO}} = \exp \frac{-tA}{M \cdot V_b \cdot B}$$

One immediate difficulty in applying such an equation is to define M. Since $O_2\overline{Hb}$ and $P\bar{c},O_2$ depend upon $\overline{COHb}$ in a rather complex fashion, it is necessary to estimate the COHb% before deciding upon an appropriate value of M. An obvious solution to this problem is to adopt an iterative computer procedure. As in the model of Goldsmith et al. (1963), a further weakness is an unduly slow estimate of the rate of response to changes in the atmospheric pressure of carbon monoxide. Under typical resting conditions, the predicted half-time of carbon monoxide elimination is 252 minutes. At equilibrium, t is equal to infinity, so that the exponential term on the right hand side of the equation becomes equal to zero. The blood carboxyhaemoglobin is then given by the sum of endogenous and exogenous terms

$$\overline{COHb}\% = \dot{V}_{CO}\frac{B}{A} + P_{I,CO}\frac{1}{A}$$

Certain assumptions are inherent in most of the equations, including (1) equilibrium of blood and body stores, (2) CO is not metabolized and is exchanged only through the lungs, (3) $O_2\overline{Hb}$ is constant in the pulmonary capillaries, and (4) ventilation, diffusion, and perfusion are uniformly distributed throughout the lungs (Klocke, 1977). Although it is relatively cumbersome, many observers consider that the model of Coburn et al. (1965) gives the most accurate representation of the facts.

The proposal of Ott and Magee (1978, cited in WHO, 1979) is similar, except that a fixed equilibrium value β of 0.5% is set for the endogenous production of carboxyhaemoglobin:

$$[\overline{COHb}\%] + \frac{\delta[\overline{COHb}\%]}{\delta t}(t) = \alpha[P_{I,CO}] + \beta$$

In this expression, when $P_{I,CO}$ is expressed in Pa, α is 1.5. The time constant for their equation is 2.49 hours. The Ott and Magee (1978) formula works quite well at the anticipated urban partial pressures of carbon monoxide (0 to 10 Pa), yielding somewhat more accurate results than methods in which carbon monoxide pressures are averaged over longer periods.

There are differences in the blood carboxyhaemoglobin concentrations predicted for a given ambient carbon monoxide pressure, depending upon the model that is accepted and the anticipated physical activity level of the individual (*see* Tables IX and X). The linear prediction equations plainly apply

Table X
INFLUENCE OF PHYSICAL ACTIVITY STATUS
UPON BLOOD CARBOXYHAEMOGLOBIN (%)*

Partial pressure of carbon monoxide (Pa)	Duration of exposure, intensity of work, and $\overline{COHb}\%$ 15 minutes S	M	H	60 minutes S	M	H	480 minutes S	M	H
0.5	0.5	0.5	0.6	0.6	0.6	0.7	0.9	0.9	0.9
1.0	0.6	0.6	0.7	0.7	0.9	1.1	1.5	1.7	1.7
2.5	0.7	0.8	1.0	1.1	1.7	2.2	3.3	4.1	4.2
5.0	0.8	1.2	1.6	1.7	3.0	4.1	6.4	8.0	8.3
20.0	1.8	3.5	5.2	5.4	11.0	15.5	24.5	31.7	32.9

*Predictions based upon model of Coburn et al., 1965 and calculations of World Health Organization, 1979 for sedentary subjects (S), subjects undertaking moderate activity (M), and subjects undertaking heavy work (H) sufficient to induce a respiratory minute volume of 30 $l \cdot min^{-1}$ and a pulmonary diffusing capacity of 60 $ml \cdot min^{-1} Torr^{-1}$.

only to the early phases of equilibration, with overestimation of blood carboxyhaemoglobin readings for long exposure periods. Discrepancies between the model of Coburn et al. (1965) and other exponential formulae are relatively slight during rest and light physical activity; it seems the best available procedure for relating carboxyhaemoglobin readings to ambient carbon monoxide pressure, duration of exposure, and physical activity status.

UPTAKE AND RELEASE OF CARBON MONOXIDE BY THE FOETUS

There is currently considerable interest in the harmful effects of carbon monoxide upon the developing foetus (Åstrup et al., 1972; MacMahon et al., 1965; Shephard, 1982a) and the newborn child (Longo, 1977). It is recognised that carbon monoxide is exchanged fairly freely across the placental membrane (Lauwerys et al., 1978), although there have been suggestions (Bisonnette et al., 1977) that some active transport mechanism becomes saturated at high carboxyhaemoglobin levels. Ethical considerations preclude the deliberate administration of carbon monoxide to pregnant women. Inferences must thus be drawn from animal experiments, occasional accidents, and women who smoke. In the last category, both carbon monoxide transport and the apparent toxicity of the gas to the foetus are influenced by other toxic constituents of cigarette smoke and socioeconomic and genetic correlates of cigarette addiction.

Nevertheless, it is becoming increasingly clear that the blockage of maternal and foetal haemoglobin by carbon monoxide (as in a heavy smoker) places the foetus at serious risk of oxygen want during the final weeks of intrauterine growth and the delivery process. The impact of differences in the chemical structure of foetal haemoglobin upon the toxicity of CO is still debated. Engel et al. (1969) observed somewhat less binding of carbon monoxide in foetal than in maternal haemoglobin, but equilibrium may not have been reached in these experiments. Longo (1977) found that the foetal pigment had a stronger affinity for carbon monoxide than its maternal counterpart, with equilibrium ratios ranging from 0.6 to 1.6 (*see* Table XI). More precisely, the foetal carboxyhaemoglobin concentration could be calculated from the formula

$$[CO\overline{Hb}]_f = [CO\overline{Hb}]_m \frac{M_f}{M_m}\left[\frac{[O_2Hb]_f \, P_{O_{2,m}}}{[O_2Hb]_m \, P_{O_{2,f}}}\right]$$

The subscripts m and f denote maternal and foetal conditions, respectively, while the other abbreviations are as in the normal Haldane formula.

The half-time for the uptake of carbon monoxide by the foetal blood is relatively long (about seven and one-half hours), so that the carboxyhaemoglobin readings for the foetus lag far behind those for the mother during acute exposure. Maternal levels are not even approximated for the first twelve hours, and up to two days are required for complete equilibration with foetal blood. The final steady-state carboxyhaemoglobin reading of the foetus is typically 10 to 15% above that of the mother (Longo, 1970, 1977; E.P. Hill et al., 1977).

The situation of the foetus is further compromised by effects of carbon monoxide upon placental metabolism and the threefold increase of endogenous carbon monoxide production in the pregnant woman.

Hill et al. (1977) have recently adapted the equation of Coburn et al. (1965) to describe the exchange of carbon monoxide between the mother and the foetus. For the mother

$$\frac{\delta[CO\overline{Hb}\%]}{\delta t} = \frac{100}{\text{Blood Volume}_m} \dot{V}_{CO,m} + \dot{D}_{L,CO}\,(P_{a,CO} - P_{\bar{c},CO}) - \dot{D}_{\bar{P},CO}(P_{\bar{m},CO} - P_{\bar{f},CO})$$

Table XI
RATIO OF FOETAL TO MATERNAL CARBOXYHAEMOGLOBIN READINGS FOR WOMEN WHO SMOKE DURING PREGNANCY

Authors	Maternal COHb%	Ratio†
Cole et al. (1972)	4.1	1.8
Haddon et al. (1961)	6.2	1.2
	3.6	0.9
Heron (1962)	6.7	0.7
Longo (1970)	6.3	0.6
Tanaka (1965)	5.7	0.9
Young and Pugh (1963)	2.0	1.2
Younoszai and Haworth (1969)	8.3	0.9

*Based on reports collected by the National Academy of Sciences, 1977.
†Note that this is the actual ratio at the time of measurement. It is not necessarily an equilibrium value.

while for the foetus

$$\frac{\delta[\overline{COHb\%}]}{\delta t} = \frac{100}{\text{Blood Volume}_f} \dot{V}_{CO,\bar{f}} + \dot{D}_{\bar{P},CO} \left(P_{\bar{m},CO} - P_{\bar{f},CO}\right)$$

The subscripts m and f here refer to maternal and foetal conditions, respectively. $\dot{V}_{CO}$ is the endogenous production of carbon monoxide, $\dot{D}_{\bar{p},CO}$ is the carbon monoxide diffusing capacity of the placenta, and $P_{\bar{m},CO}$ and $P_{\bar{f},CO}$ are the mean partial pressures of carbon monoxide in the maternal and placental capillaries.

Under steady-state conditions, foetal carbon monoxide production contributes about 3% of the carboxyhaemoglobin burden of the pregnant woman.

Chapter 4

HEALTH EFFECTS OF CARBON MONOXIDE EXPOSURE

There is little question that high concentrations of carbon monoxide are toxic to man, but there is much less certainty concerning the minimum ambient pressures at which adverse health effects begin. Careful study is required since the costs of total elimination of carbon monoxide from the urban and the industrial environment are prohibitive.

This chapter looks at techniques used in establishing threshold doses of carbon monoxide (animal experimentation, epidemiological surveys, rapid rebreathing of carbon monoxide, and deliberate chamber exposure to the gas). Particular attention is concentrated upon cardiorespiratory changes (an increase of heart rate and ventilation, a deterioration of maximum performance) and disturbances of central nervous function (vision, hearing, cerebration, psychomotor performance, and vehicle operation). The acute response to occasional more severe exposures is discussed, and the possible cumulative effects of repeated chronic exposure to carbon monoxide are examined.

TECHNIQUES OF INVESTIGATING HEALTH EFFECTS

Animal Models

There are several advantages to the use of animal models in elucidating the health effects of carbon monoxide. In particular, it is possible to test in an animal responses to concentrations of carbon monoxide that would be dangerous to man, experimental exposures can be continued over much of the animal's lifespan, and specimens of various tissues can subsequently be excised for detailed microscopic and histochemical examination.

On the other hand, there are substantial species differences in a number of factors influencing carbon monoxide toxicity, including respiratory minute volume, pulmonary diffusing capacity, blood volume, the shape of the oxygen dissociation curve, and the carbon monoxide affinity of haemoglobin

(Klemisch et al., 1975). It is also quite difficult to persuade many animals to undertake maximal exercise, and it is even more difficult to detect a small degradation in their psychomotor performance.

Epidemiological Surveys

It is difficult to persuade even healthy subjects to inhale toxic gases, and few of those with cardiac or respiratory disease are willing volunteers for carbon monoxide inhalation experiments. As with other types of air pollution, there has thus been substantial interest in the epidemiology of incidental industrial and urban exposures.

The pitfalls of epidemiological research are well known (Shephard, 1974). In many cities, the usually encountered concentrations of carbon monoxide are below the probable threshold for initiation of either a physiological or a pathological response. Moreover, the measured partial pressures of carbon monoxide refer to a single monitoring site or at best to a limited sampling network, so that the air quality statistics are often unrepresentative of individual exposure patterns. If correlations are found between regional peaks of carbon monoxide exposure and poor health, the association is not necessarily causal. High concentrations of carbon monoxide may occur concurrently with other industrial or vehicular pollutants and may be related to specific climatic conditions such as a bout of cold weather. People living in heavily polluted sections of a city may also be at a socioeconomic disadvantage, economic pressures and social alienation leading to an adverse life-style and a poor health experience. Incidental exposure may even fail to show clear peaks of pollution; it then becomes difficult to establish reference points that can be correlated with subsequent responses. Fortunately, with many of the probable health effects of carbon monoxide, there is less need to allow for a lag period than when dealing with other pollutants (Shephard et al., 1960). Finally, for about 35 to 40% of most adult populations, self administration of carbon monoxide in cigarette smoke far outweighs incidental industrial or urban exposure (Rode et al., 1972).

Rapid Rebreathing Exposure

Godin and Shephard (1972) suggested that the rapid rebreathing technique used to estimate blood carboxyhaemoglobin levels could also be exploited for administration of carbon monoxide. A known amount (40 to 100 ml) of pure carbon monoxide was added to a two-litre bag of oxygen, expired carbon dioxide being absorbed in a soda lime canister. After two minutes of rebreathing, the increment of carboxyhaemoglobin in a nonsmoker ranged from about 2.4% with 40 ml to 5.4% with 100 ml of carbon monoxide. For control experiments, it was a simple matter to add 40 to 100 ml of air to the bag of oxygen.

The approach proposed is extremely safe since it is impossible for the intended dose of carbon monoxide to be exceeded. Further, the timing of the rise in blood carboxyhaemoglobin is known with considerable precision. The pattern of exposure is somewhat analogous to that encountered at a particularly dangerous phase of a smelting operation or during the smoking of one or more cigarettes. The main disadvantage is that the initial blood concentrations of carbon monoxide exceed equilibrium value, with redistribution of CO occurring between the various body compartments while the physiological effects of a given dose are being observed.

Chamber Exposure

There is no fundamental reason why volunteers should not be exposed to carbon monoxide mixtures using the large gas chambers still available at some military establishments; parties of physically active subjects could then be tested simultaneously. However, for reasons of expense, university-based investigators have usually conducted their experiments upon single subjects sitting in very small exposure booths (O'Donnell et al., 1971; Theodore et al., 1971; Wright and Shephard, 1978c). If the experiment is not controlled, any observed changes of psychomotor performance may then be due to sensory isolation rather than carbon monoxide poisoning. Even if controlled experiments are arranged, responses may reflect an interaction between carbon monoxide and sensory isolation rather than an effect of carbon monoxide that would be observed in a more normal environment.

The main advantages of a chamber exposure are that one can mimic the long duration/low concentration exposures likely in the urban environment and the subjects are unaware when the exposure has commenced.

CARDIORESPIRATORY EFFECTS OF CARBON MONOXIDE

Quite modest increases of blood carboxyhaemoglobin level can influence both resting cardiorespiratory variables and the response to various intensities of exercise. Although such changes are relatively easily measured, they have little practical significance for the health of the normal adult. Likewise, a modest increase of blood carboxyhaemoglobin may do no more than cause a minor decrement of athletic performance. Nevertheless, the blood supply to the heart muscle is marginal for the demands of vigorous effort, particularly in an older person with some atherosclerosis, and there is little scope for development of a cardiac oxygen debt. Carbon monoxide can thus impair the performance of the heart. Electrocardiographic abnormalities and an impairment of myocardial contractility have been demonstrated in animal experiments, but the concentrations of carbon monoxide used have been much larger than those likely in urban or industrial exposures to this gas.

Resting Conditions

Studies of the carotid body have suggested that carbon monoxide poisoning should stimulate the cardiac and respiratory control centres of the medulla.

Permutt and Farhi (1969) estimated that a 9% increase of blood carboxyhaemoglobin level would lower the venous (and thus the tissue) oxygen pressure by 0.5 to 0.8 kPa in tissues with a wide arteriovenous oxygen difference; as discussed in the previous chapter, this reflects both binding of carbon monoxide to the haemoglobin molecule and an alteration in the shape of the oxygen dissociation curve. Unfortunately, the precise arteriovenous oxygen difference for the carotid body is unknown; it probably averages only about 5 $ml \cdot l^{-1}$, varying with the activity of the efferent nerve supply to this organ. Comroe (1965) calculated that even if there were a 50% increase of carboxyhaemoglobin level, the end-capillary oxygen pressure of the carotid body would only drop from 11.7 to 10.6 kPa. Assuming no change in the arterial carbon dioxide pressure (isocapnic conditions), a 1 kPa decrease of oxygen pressure increases ventilation by only about 0.36 $l \cdot min^{-1}$ (Loeschke and Gertz, 1958). A more modest 5% increase of carboxyhaemoglobin would reduce the carotid body oxygen pressure by no more than 0.1 kPa, with a barely detectable 0.04 $l \cdot min^{-1}$ change of respiratory minute volume.

Experiments on cats have allowed measurement of impulse traffic in the afferent nerves from the carotid chemoreceptors. Substantial poisoning (carboxyhaemoglobin readings of 18 to 40%) cause a two- to fivefold increase of nerve impulses (Duke et al., 1952; Mills and Edwards, 1968; Paintal, 1967). The response is observed despite some increase of arterial oxygen pressure. It is not attributable to hypotension or a direct effect of carbon monoxide upon the cytochrome oxidase system of the carotid bodies, and by exclusion it has been attributed to the onset of tissue hypoxia. Measurements of brain and skeletal muscle oxygen pressures have been made during exposure to 8 Pa of carbon monoxide; the average response to this level of poisoning was a 10.6 Pa decrease of tissue oxygen pressure (Weiss and Cohen, 1974). Likewise, puffs of carbon monoxide sufficient to induce blood carboxyhaemoglobin readings of 4 to 7% reduced the oxygen pressure within cardiac muscle by as much as 1 to 4 Torr (0.13 to 0.53 kPa; Rink, 1978).

When human volunteers have been exposed to mild doses of carbon monoxide, changes of heart rate, myocardial contractility, and respiratory minute volume have been small and equivocal (Horvath et al., 1971; Klausen et al., 1968; Shephard, 1972; Weiss and Cohen, 1974; Raeder et al., 1979; see Table XII). Ayres et al. (1970) found an increase of cardiac output from 5.0 to 5.6 $l \cdot min^{-1}$ and of respiratory minute volume from 6.9 to 8.6 $l \cdot min^{-1}$ at a carboxyhaemoglobin reading of 6 to 12%. More severe carbon monoxide poisoning is reported as causing hyperventilation in both goats (Santiago and Edelman, 1976) and man (Leathart, 1962). The mechanism is apparently tissue acidosis rather than chemoreceptor stimulation. Daly et al. (1954) pointed out that a 75% reduction of oxygen flux due to carbon monoxide exposure caused no increase of impulse traffic in the sensory nerves arising from the carotid chemoreceptors. This is a little surprising since the cytochrome oxidase system is involved in signal generation; presumably, the affinity of cytochrome oxidase for oxygen is sufficiently high relative to that for carbon monoxide that function is not disturbed (Neil and Joels, 1963; Horvath et al., 1971). Although a two- to fivefold increase of apparent impulse traffic has been demonstrated in some animal experiments, quite high carboxyhaemoglobin readings (18 to 40%) were reached. Comroe (1965) estimated that a 50% saturation of the blood with carbon monoxide would reduce the end-capillary oxygen pressure in the carotid body sinusoids by no more than 8 Torr, and this should increase ventilatory drive by less than 5% (Loeschke and Gertz, 1958). It is thus plain that a 2 to 5% increase of carboxyhaemoglobin is unlikely to have a measurable effect upon the cardiovascular system. There is little or no stimulation of respiration, even with severe poisoning; a further factor contributing to this finding is a depression of the medullary respiratory control centres by hypoxia (Chiodi, 1941).

Inhalation of carbon monoxide commonly increases the resting coronary blood flow. Adams et al. (1973) found a 13% increase with a 4% rise of blood carboxyhaemoglobin and a 54% increase at 20% carboxyhaemoglobin. If compensation is incomplete, there may be a reduction of intraventricular oxygen pressure (Rink, 1978), but there is usually little impairment of myocardial contractility (Raeder et al., 1979). Asmussen and Vinter-Paulsen (1949) even observed a small increase of cardiac output with severe carbon monoxide poisoning. Davies and Smith (1980) exposed volunteers to known concentrations of carbon monoxide for eighteen days; twelve-lead electrocardiograms showed ST changes in one of fifteen subjects and P wave changes in three of fifteen at 1.5 Pa, P wave changes in six of fifteen subjects at 5 Pa, and abnormalities in seven of ten subjects at 7.5 Pa of carbon monoxide.

Table XII

AVERAGE HEART RATE FOR 30 MINUTES BEFORE AND 30 MINUTES AFTER TWO-MINUTE REBREATHING OF CARBON MONOXIDE MIXTURE (2.4% OR 5.4% INCREASE OF CARBOXYHAEMOGLOBIN)*

Experimental exposure			Control exposure		
ΔCOHb%	Heart Rate (beats/min) Initial	Final	ΔCOHb%	Heart Rate (beats/min) Initial	Final
+2.4	68.2	68.0	−0.02	65.3	64.2
+5.4	63.6	60.5	−0.02	61.9	60.4

*Based on data of Shephard, 1972.

Submaximal Exercise

The early observations of Haldane suggesting difficulty in movement as a result of severe poisonings are discussed below. Chiodi et al. (1941) have also commented that with a 40 to 50% blood carboxyhaemoglobin reading, subjects had difficulty undertaking tasks that required even moderate effort. Lesser exposures may modify physiological responses to submaximum exercise through chemoreceptor stimulation, as discussed above. In addition, interference with oxygen transport to the working muscles may increase lactic acid formation and thus the respiratory response to a given intensity of exercise (Calverley et al., 1979). The extent of cardiorespiratory responses depends upon both the intensity of exercise and the blood carboxyhaemoglobin level; in general the higher the fraction of maximum oxygen transport that is deployed, the lower the carboxyhaemoglobin reading needed to increase respiratory minute volume and heart rate.

Several authors (Ekblom and Huot, 1972; Klausen et al., 1968; Pirnay et al., 1971; Vogel et al., 1972; Wagner et al., 1978) have shown that moderate carbon monoxide exposure has no effect upon the oxygen cost of a given submaximal power output. Pitts and Pace (1947) noted that during mild exercise, the heart rate was unchanged at a carboxyhaemoglobin level of 6% but was increased at 13% carboxyhaemoglobin. On the other hand, Wagner et al. (1978) found that carboxyhaemoglobin readings of 6 to 21% had no effect upon the cardiovascular and respiratory responses of dogs to exercise. Likewise, Asmussen and Chiodi (1941) and Apthorp et al. (1958) observed that neither respiratory minute volume nor respiratory rate responses to exercise were changed, even at carboxyhaemoglobin readings in excess of 30%; however, the steady-state heart rate was increased in these studies, with implications for the tolerance of sustained physical activity.

Many published reports have discussed fairly severe exposures to carbon monoxide. Chevalier et al. (1966) noted an increase of oxygen debt in smokers (4% carboxyhaemoglobin) relative to nonsmokers; however, this may have reflected a long-term impairment of muscle blood supply in the smokers or a response to some other toxic constituent of cigarette smoke. Gliner et al. (1975) had their subjects work for a relatively long period (four hours at 35% of maximum oxygen intake) while exposed to a carbon monoxide partial pressure of 5 Pa. Under these conditions, the final blood carboxyhaemoglobin values were surprisingly high (10.3% and 13.2% in nonsmokers and smokers respectively). The heart rate response to the required intensity of exercise was higher in the CO mixture than when breathing air, and a decreased tolerance of heavy work was shown by the fact that tachycardia began to develop after two hours of effort. Seppanen (1977) induced a 9.9% carboxyhaemoglobin and found decreases in the power output at heart rate of 130, 150, and 170 beats$\cdot$min^{-1}, while Brinkhouse (1977) commented on a significant reduction in the time to reach a target heart rate during treadmill exercise at 5 or 10 Pa of carbon monoxide. At moderate altitude (1610 metres), the CO-induced decrease of working capacity is much as at sea level, but the anerobic threshold is diminished, and probably for this reason the respiratory minute volume at a given power output is increased (Weiser et al., 1979).

Maximal Exercise

Changes of maximal performance can be detected with quite modest exposures to carbon monoxide (Aronow and Cassidy, 1975; Brinkhouse, 1977). Furthermore, the relationship between increase of blood carboxyhaemoglobin level and decrease of maximum oxygen intake is essentially linear, following the equation (Coburn, 1977)

$$\dot{V}O_2\,(max)\% = 0.91\,[CO\overline{Hb}] + 2.2$$

There is no evidence that the magnitude of response is modified by a specific pattern of carbon monoxide exposure (high partial pressure/brief exposure versus low pressure/long exposure), but there is (surprisingly) sometimes less effect on smokers than on nonsmokers. Most of the subjects tested to date have been young adults, and there is a need to extend observations to older people. There have been attempts to demonstrate a "threshold" dose, but the lack of a significant decrease in maximum oxygen transport at carboxyhaemoglobin levels of less than 3 to 4% seems due to errors in the measurement of maximum oxygen intake and carboxyhaemoglobin rather than the absence of an effect; extrapolation of the linear relationship confirms that the effect upon oxygen transport persists down to the lowest carboxyhaemoglobin levels.

The heart rate, respiratory rate, and respiratory minute volume responses to maximum exercise remain unchanged by moderate carbon monoxide exposure (Ekblom and Huot, 1972; Maugh, 1972; Vogel et al., 1972), but there is a decline of endurance time (Brinkhouse, 1977) and of maximum oxygen intake that is approximately proportional to the increase of blood carboxyhaemoglobin (Vogel and Gleser, 1972).

Carboxyhaemoglobin formation is the main reason for the loss of aerobic power, reflecting both carbon monoxide binding and the shift of the oxygen dissociation curve. Clark and Coburn (1975) postulated an additional effect from intracellular responses to carbon monoxide, but there is litle evidence that "bolus" administration of the gas produces a larger response, as would be expected with such a mode of action. Raven et al. (1974) and Horvath et al. (1975) suggested the decline of aerobic power became nonlinear at a carboxyhaemoglobin reading of less than 6%, but the published data is consistent with a linear relationship continuing to the lowest carboxyhaemoglobin values. The experiments from the Santa Barbara laboratory were supposedly double-blind in nature, but in practice subjects identified experiments where they had inhaled high concentrations of carbon monoxide, commenting that the lower extremitites felt heavy and exercise seemed more difficult to perform.

Several authors have noted decrements in the endurance time for near-maximal exercise. Ekblom and Huot (1972) reported a 38% decrease of endurance at a carboxyhaemoglobin level of 7%, while Drinkwater et al. (1974) and Raven et al. (1974) both commented on a shortening of working time at a carbon monoxide partial pressure of 5 Pa (rise of carboxyhaemoglobin 2.1% in nonsmokers and 2.6 to 3.2% in smokers).

Macmillan (1969) and Goldsmith (1970) suggested a carbon monoxide pressure of 3 Pa was sufficient to reduce competitive swimming performance. Since the results of swimming meets are timed to a small fraction of a second, detection of a response to low CO pressures is quite likely. However, carboxyhaemoglobin was not determined, and as the carbon monoxide was emitted by vehicular traffic, other pollutants may have contributed to the observed decrement of swimming speeds.

EFFECTS OF CARBON MONOXIDE UPON THE CENTRAL NERVOUS SYSTEM

General Considerations

Reports of impaired auditory duration discrimination at low levels of carbon monoxide exposure (2.0 to 2.5% carboxyhaemoglobin) have stimulated a vigorous search for changes of psychomotor performance that may result from normally encountered levels of carbon monoxide (Stewart, 1976). To date, several factors have contributed to the appearance of conflicting and rather unsatisfactory data.

Animal experiments necessarily have only limited value. Patterns of uptake, distribution, and elimination of carbon monoxide differ from those anticipated in man, and it is also arguable that the animal brain differs qualitatively or quantitatively in its responses to carbon monoxide-induced hypoxia. There is generally difficulty in carrying out and interpreting psychomotor tests on species other than man. What, for instance, are the implications of a reduction in the number of bar presses made by white rats after two hours' exposure to carbon monoxide at a pressure of 20 Pa (Ator et al., 1976; Laties and Merigan, 1979; Goldberg and Chappell, 1967)? Has there been modification of a learned response, a reversal of a previously learned discrimination, loss of appetite, or change in some other biological function? Nevertheless, some hypotheses have emerged from animal experiments. Beard and Wertheim (1967) perceived a derangement of time sense when rats were exposed to carbon monoxide, while Back and Dominiguez (1968) and Back (1969) used *Macaca mulatta* monkeys to investigate the effects of long-term carbon monoxide exposure upon responses to simple auditory and visual stimuli; the results suggested no impairment of learned responses with carboxyhaemoglobin values of up to

30% (Theodore et al., 1971). Animal experiments have also allowed the direct measurement of cerebral blood flow; this seems to be affected more by an accumulation of carbon monoxide than by exposure to an equivalent oxygen lack (Traystman and Fitzgerald, 1977). Even a 2% increase of the blood carboxyhaemoglobin is sufficient to impair oxygen delivery to the brain (Zorn, 1972), with a reduction of local oxygen pressure (but not of oxygen usage) in the cerebral tissue. Dogs do not show any compensatory increase of cerebral blood flow below a carboxyhaemoglobin level of 20% (Haggendal et al., 1966). With more severe exposures, the increase of blood flow is sufficient to avert a further widening of cerebral arteriovenous oxygen difference. Apparently, the augmentation of blood flow also prevents the development of cerebral hypoxia, since there is no increase of lactate concentrations in jugular vein blood.

The psychomotor tests used in man often have prolonged learning schedules and lack precision. Performance is commonly affected by arousal of the subject, leading to interactions between personality, environmental stress, and test score. Because the typical arousal curve has an inverted U-shape, a small dose of carbon monoxide may improve test performance while a larger dose has an adverse effect. Depending upon personality, different individuals respond to the same dose of carbon monoxide in opposite ways. The behaviour of any given subject is also reversed if other environmental circumstances are altered. The experiment itself can serve as an arousing stimulus, but on the other hand the conditions of gas exposure (often confinement to a small, closed chamber) can lead to a loss of vigilance. Many of the simpler psychomotor tests leave a substantial reserve of mental function untapped, and this reserve is decreased by carbon monoxide exposure. The true extent of functional impairment may thus pass unnoticed until an attempt is made to complete what had previously seemed a simple manoeuvre under the added stress of an emergency. This limitation of conventional testing has obvious relevance in considering the impact of carbon monoxide exposure upon the risk of vehicular and other accidents.

Several respected authors found no significant central nervous effects of carbon monoxide, even at very high blood levels. Forbes et al. (1937) reported no alteration of reaction time, binocular vision, and eye-hand coordination despite carboxyhaemoglobin readings of 30%. Again, Stewart et al. (1970) observed no change of hand and foot "reaction times," hand-steadiness, manual dexterity, visual acuity, depth perception, and time estimation with carboxyhaemoglobin levels of 11 to 13%, while O'Donnell et al. (1971) detected no changes of critical flicker fusion frequency, time estimation, and mental arithmetic ability at carboxyhaemoglobin levels of up to 12.7%.

In contrast, McFarland et al. (1944) and Halperin et al. (1959) both noted changes of visual intensity discrimination at carboxyhaemoglobin values of 3 to 5%. Likewise, Schulte (1963) observed a deterioration of choice discrimination (for colour and letters) at a carboxyhaemoglobin level of less than 5%, with parallel changes in scores for such cognitive tasks as mental arithmetic, plural noun underlining, and ± deletion. Horvath et al. (1971) saw a worsening of visual vigilance in nonsmokers at an average carboxyhaemoglobin of 6.6%, while Ramsey (1972) described a small deterioration in reaction times at a value of 5%.

While many authors seem willing to accept that there is some loss of psychomotor function at 3 to 4% carboxyhaemoglobin (see Table XIII), the loss of auditory discrimination observed by Beard and Wertheim (1967) at an estimated 2 to 3% carboxyhaemoglobin remains controversial (O'Donnell et al., 1970; Stewart et al., 1970; Weir et al., 1973; Otto et al., 1978; Wright and Shephard, 1978; Beard and Grandstaff, 1975). In view of the discordant results, the experimental evidence is reviewed in more detail below.

Table XIII
SUMMARY OF CHANGES IN CENTRAL NERVOUS FUNCTION REPORTED FOLLOWING CARBON MONOXIDE EXPOSURE

Psychomotor function	Threshold COHb%	Comment	Author
Vigilance task	1.8	? Exaggerated by sensory isolation. Effect reversed at highest dose	Beard and Grandstaff (1972)
Vigilance task	2.3-3.1	Effect disappeared with continuation of experiment	Fodor and Winneke (1972)
Vigilance task (weaker tones)	?3.0	COHb% not measured	Groll-Knapp et al. (1972)
Vigilance (brighter pulse)	5.0		Horvath et al. (1971)
Time estimation	1.8	Sensory isolation	Beard and Grandstaff (1972)
Auditory duration discrimination	2.5-4.0	Sensory isolation COHb% not measured	Beard and Wertheim (1967)
Visual intensity discrimination	3.0	Sensory isolation	Beard and Grandstaff (1970a)
Visual acuity	4.5		McFarland et al. (1974)
Tracking task	3.0	? Statistically unreliable	Mikulka et al. (1971)
Choice reaction time	3.5-4.0	Many negative results for simple reaction time	Ramsey (1972)
Mental performance, finger dexterity	7.2		Bender et al. (1971)
Hand/eye coordination + peripheral vision	5.0	High tracking rate	Putz et al. (1976 and 1979).

Visual Responses

Vision is one of the earliest functions to be disturbed when cerebral hypoxia is induced by a reduction of the inspired oxygen pressure. It is thus not surprising that many investigations of carbon monoxide toxicity have concentrated upon visual responses. Properties to be examined include spatial discrimination (visual acuity), intensity discrimination (differential brightness sensitivity), temporal discrimination (critical flicker fusion frequency), and adaptation to low intensities of illumination.

McFarland and Halperin (1940) found a significant rightward displacement of the visual acuity curve along the illumination axis at a simulated altitude of 3050 metres (inspired oxygen pressure ~ 14.1 kPa). Likewise, when carbon monoxide was administered by means of a facemask, decreases of visual acuity were seen at carboxyhaemoglobin readings as low as 4.5%, with recovery lagging behind carbon monoxide elimination (see Figure 7; McFarland et al., 1944). Beard and Grandstaff (1970a) also reported a 5% reduction of visual acuity with increases of brightness threshold after twenty-seven minutes' exposure to 5 Pa of carbon monoxide (a 3% carboxyhaemoglobin reading being estimated from alveolar gas samples); larger changes (16 and 32% decrements of acuity) were seen at estimated carboxyhaemoglobin levels of 6 and 10%, respectively. In contrast, Stewart et al. (1970) saw no impairment of visual acuity when eighteen healthy men were exposed to 10 Pa of carbon monoxide for eight hours (11 to 13% carboxyhaemoglobin); the only rather

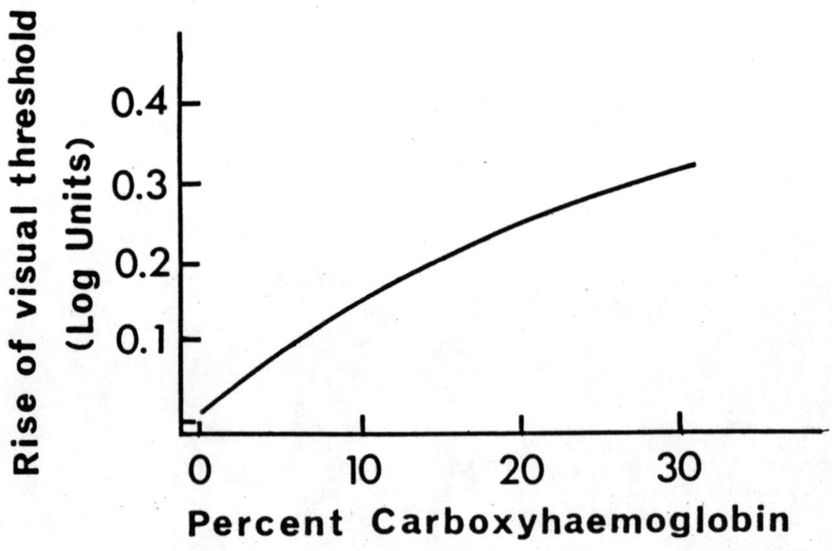

Figure 7. Influence of an increase of blood carboxyhaemoglobin upon the visual threshold. Based on data of McFarland et al. (1944).

unconvincing explanation that has yet been advanced for their discordant findings is a failure to use the precise psychophysical testing technique of limits (Corso, 1967).

The visual discriminometer of Crozier and Holway (1939) allows the testing of differential light sensitivity in a dim and uniformly illuminated field. McFarland et al. (1944) and Halperin et al. (1959) observed a loss of discrimination with carbon monoxide exposure, corresponding roughly in magnitude with the response to an equal reduction of arterial oxygen saturation induced by the breathing of low oxygen mixtures. The apparatus was suggested to be capable of detecting the effect of smoking a single cigarette (a 2% increase of blood carboxyhaemoglobin). These observations were confirmed by Beard and Grandstaff (1970a), who found a statistically significant increase of the intensity discrimination threshold with forty-nine minutes of exposure to 5 Pa of carbon monoxide (alveolar samples equivalent to 3% carboxyhaemoglobin); a response was also reported with seventeen minutes' exposure to 15 or 25 Pa of carbon monoxide. Again, there are puzzling discrepancies between these findings and more recent reports. Ramsey (1972, 1973) found no change of either brightness discrimination or depth perception with a carboxyhaemoglobin of 9 to 16%, while Weir et al. (1973) saw no change in the brightness matching of three pairs of coloured lights at 7 or 14% carboxyhaemoglobin.

Most authors have found that the critical flicker fusion frequency (Landis, 1954) is unchanged by commonly encountered carbon monoxide exposures (Lilienthal and Fugitt, 1946; Vollmer et al., 1946; Guest et al., 1970; O'Donnell et al., 1971a; Ramsey, 1972; Winneke, 1974; Weber et al., 1975). The one positive report was that of Seppanen et al. (1977). Lilienthal and Fugitt (1946) also observed an effect when 5 to 9% carboxyhaemoglobin was combined with exposure to an altitude of 1500 to 1800 metres, but Vollmer et al. (1946) had negative findings when a 22% carboxyhaemoglobin was produced at 4700 metres simulated altitude.

The threshold of illumination for dark-adapted subjects was increased in seven of nine subjects at a carboxyhaemoglobin level of 10% (Abramson and Heyman, 1944). Nevertheless, there was great individual variation of response. Intersubject and intrasubject variability may have contributed to subsequent negative findings. Weir et al. (1973) reported no change at 7 or 14% carboxyhaemoglobin, while McFarland (1973) saw no response at 11 or 17% carboxyhaemoglobin. Wright et al. (1973) observed no change of night vision, glare vision, glare recovery, or depth perception with an acute 3.4% increase of carboxyhaemoglobin values, while Luria and McKay (1979) found no change of scotopic sensitivity or reaction time with three hours' exposure to 22 Pa of carbon monoxide.

There have been suggestions that carbon monoxide exposure changes the visual evoked response of rats (Xintaras et al., 1966) and cats (Ingenito, 1979), although it is uncertain whether this reflects hypoxia or a more direct effect of carbon monoxide upon retinal metabolism (Ingenito and Durlacher, 1979). Humans are apparently unaffected until a carboxyhaemoglobin of 20 to 28% is reached (Hosko, 1970; Sprout et al., 1975; Luria and McKay, 1979); indeed, Dinman (1969) reported no change of latency or voltage with a carboxyhaemoglobin of 37%, while the spontaneous electrical activity of the brain was unchanged at 33% COHb (Hosko, 1970).

Auditory system

The auditory variables that can be examined are analogous to those for the visual system. Some, such as intensity threshold, intensity discrimination, frequency discrimination, and sound localization have not been tested in relation to carbon monoxide toxicity. However, there are published reports on auditory duration discrimination and auditory flutter fusion.

Beard and Wertheim (1967) required subjects to compare the duration of pure 1000 Hz tones. The mean percentage of correct responses decreased from 78 to 68% after 90 minutes' exposure to 5 Pa of carbon monoxide (likely to have induced carboxyhaemoglobin readings of 2.5 to 4.0%). Critics of the experiment note that subjects were isolated in an audiometric booth with relatively infrequent testing bouts and had spent thirty minutes in the booth even before the carbon monoxide exposure began; for this reason the tests measured vigilance as much as time discrimination. Beard, working with Grandstaff (1970b), was unable to replicate this result using paired *visual* signals despite exposure of subjects to carbon monoxide pressures as high as 25 Pa. In the second experiment, Beard and Grandstaff (1970b) chose to avoid the range ± 125 milliseconds, where discrimination became difficult (Wright and Shephard, 1978c).

O'Donnell et al. (1971) and Theodore et al. (1971) argue that the effect reported by Beard and Wertheim (1967) was due to sensory isolation rather than to carbon monoxide poisoning. Their subjects were exposed to 7.5 or 15.0 Pa of carbon monoxide for 7.5 hours, with final carboxyhaemoglobin readings of 5.9 and 12.7% respectively. Details of the rate of signal presentation were not given, and the experimental design (an incomplete Latin Square) precluded comparisons between carbon monoxide exposure and air breathing; nevertheless, a comparison of the results for 5.9 and 12.7% carboxyhaemoglobin suggested that carbon monoxide exposure had no effect upon the auditory duration discrimination threshold. Stewart et al. (1973) exposed subjects in groups of five, allowing some interindividual interaction for part of each hour. Problems of sensory isolation were thus avoided. Using a double-blind protocol, they saw no change of performance with exposures to 5, 10, 20, or 50 Pa of carbon monoxide for 2.5 or 5.0 hours. Nevertheless, certain differences from the Beard and Wertheim protocol must be stressed. The second signal followed the first by 1.5 rather than 0.5 seconds, with a correspondingly greater potential for fading of the auditory image (Corso, 1967; Massaro and Kahn, 1973; Plomp, 1964). Moreover, discrimination was tested only once per hour rather than every twenty minutes, and the performance of other tasks during the intervening period probably helped to sustain arousal (Beard and Grandstaff, 1972; Fodor and Winneke, 1972), with beneficial consequences for the vigilance of their subjects (Wright, 1971). In support of this view, Stewart et al. (1973) noted that nine subjects who were confined one at a time to an audiometric booth showed a significant (2.9%) decrement of signal detection at 9.7% carboxyhaemoglobin. Wright and Shephard (1978c) repeated Beard and Wertheim's 1967 experiment, giving a rapid, bolus-type rebreathing exposure to carbon monoxide. The subjects sat in a quiet room one at a time with a solitary observer; again, the most important variable influencing test scores was time on the task. The simplest analysis of their data indicated a significant increase of mean errors at blood carboxyhaemoglobin levels as low as 3.2%, the dose/response line (see Figure 8) running in between the previously published lines of Beard and Wertheim (1967) and Stewart et al. (1973).

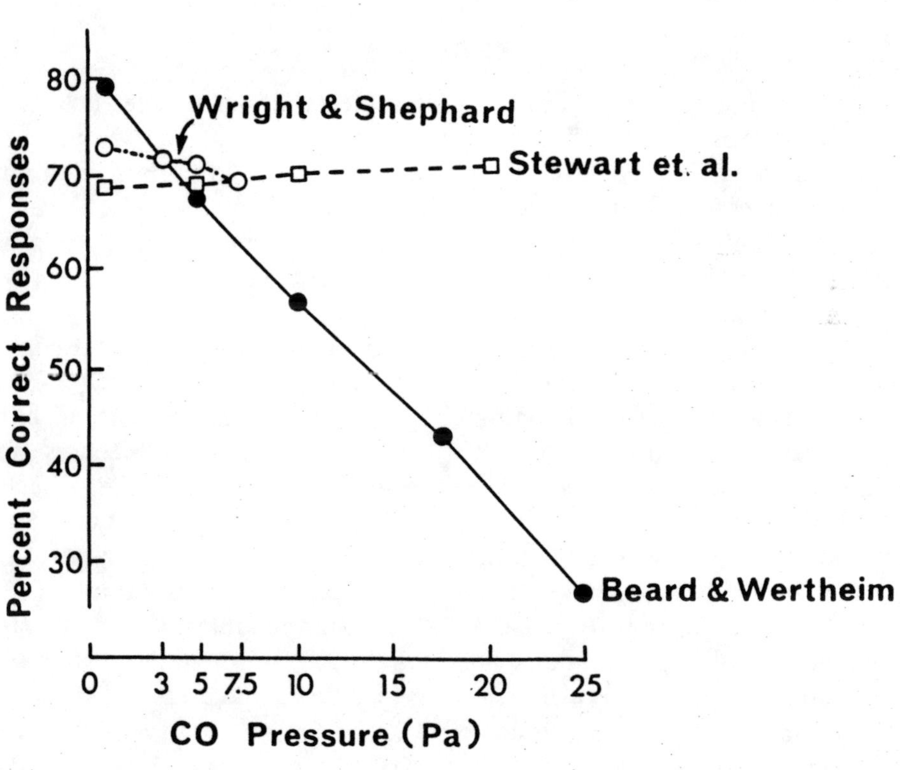

Figure 8. The dose/response line for an auditory discrimination task. Based on published results of Beard and Wertheim, 1967, Stewart et al., 1973, and Wright and Shephard, 1978c, for subjects exposed to carbon monoxide.

Unfortunately, the conclusion that is drawn depends not only upon test conditions but also upon the method of data analysis. Beard and Wertheim (1967) used the method of constant stimulus differences, which most psychophysiologists regard as a more appropriate measure of auditory duration discrimination than the simple analysis of percentage correct responses. When Wright and Shephard (1978c) applied the psychophysical method of analysis to their data, the previously observed small effect disappeared, and indeed no deterioration of auditory duration discrimination could be demonstrated to a carboxyhaemoglobin reading of 8%. Otto et al. (1979) likewise observed no responses to carboxyhaemoglobin levels of 7.8%. Auditory flutter fusion frequency is apparently unchanged even at blood carboxyhaemoglobin levels of 10% (Guest et al., 1970).

SENSORY PROCESSING

The responses of the central nervous system depend upon a vigilance in detecting sensory stimuli, a capacity for time estimation, learning and memory, mental and cognitive performance, and choice reaction time.

Vigilance

A typical vigilance test examines the subject's ability to detect a small change of environment (such as a brief blip on a radar screen) occurring at an unpredictable frequency. Quite modest exposures to carbon monoxide have sometimes reduced auditory and/or visual vigilance. Lewis et al. (1970) noted an effect when subjects breathed air pumped into a stationary car from the side of a busy road; CO pressures were apparently above 1 Pa for 40% of the time and over 3 Pa for 4% of the time. Groll-Knapp et al. (1972) required subjects of unspecified smoking habit to detect slightly weaker tones over ninety minutes' exposure to carbon monoxide (5, 10, or 15 Pa); the blood carboxyhaemoglobin was not measured but probably reached values of 3.0, 5.4, and 7.6% respectively. Such exposures were sufficient to induce a progressive deterioration of performance, the respective number of missed signals being 26, 35, 40 and 44 for the four conditions. There were also changes in associated brain wave potentials. However, other workers from the same laboratory were unable to replicate these observations (Haider et al., 1975). Fodor and Winneke (1972) reported a substantial (20 to 30%) decrease of signal detection and a 10% increase in response latency after 80 to 110 minutes of exposure to carbon monoxide (5 Pa partial pressure, equivalent to a blood carboxyhaemoglobin of 2.3 to 3.1%). However, the effect disappeared as the exposure continued, suggesting that some compensation may have developed. Winneke (1974) carried out further studies at carbon monoxide pressures of 5 and 10 Pa. It is likely that the final carboxyhaemoglobin level reached 9%, but there were no significant changes of signal detection. Putz et al. (1976 and 1979) had their subjects carry out an auditory vigilance task over four hours' exposure to 3.5 or 7.0 Pa of carbon monoxide, the latter generating carboxyhaemoglobin readings of 5%. Such exposures led to no change of response times, number of misses, false alarms, or percentage of correct detections, but there were apparently changes in the auditory evoked cerebral action potential at both exposure levels.

Horvath et al. (1971) required their subjects to detect a slightly brighter light pulse. After breathing 11.1 Pa carbon monoxide for sixty minutes (average carboxyhaemoglobin about 5%), a significantly smaller number of signals was detected (55 to 60%, versus 70% after thirty minutes of exposure). However, the effect did not persist if the arousal of the subjects was increased by boosting the rate of signal presentation from two to ten per three-minute interval. Beard and Grandstaff (1972) had their subjects report shorter light pulses while exposed to carbon monoxide in a small audiometric booth. Signal detection deteriorated significantly (on the basis of a one-tailed t test) at the two lowest carbon monoxide concentrations (1.8 and 5.2% carboxyhaemoglobin) but returned almost to normal at 7.5% carboxyhaemoglobin (this paradoxical finding being reminiscent of the data of Fodor and Winneke, 1972). It was argued that at the two lower doses of carbon monoxide, the subjects showed not only a decreased sensitivity to signals but also increased caution in responding.

As might be anticipated, the maintenance of arousal has a critical influence upon the outcome of vigilance studies. Winneke et al. (1976) were unable to replicate the findings of Horvath et al. (1971) under conditions that avoided sensory deprivation. Likewise, other workers in Horvath's laboratory (Christensen et al., 1977) failed to find any deterioration of visual vigilance at 6% carboxyhaemoglobin when exposures were conducted in a clear plastic chamber, while Benignus and Otto (1977) found no deterioration in a numeric monitoring task with carboxyhaemoglobin levels of up to 12.6%.

Time Estimation

Psychophysiologists distinguish a time duration sense (which depends upon an interval timing ability) from a true time or orientation sense (Sollberger, 1967).

Beard (1969) had subjects estimate ten- and thirty-second intervals, using a push button. Exposure to 5 Pa of carbon monoxide for seventy-five minutes (estimated carboxyhaemoglobin of less than 3%) was without effect upon the ten-second estimates, but it impaired judgements of the thirty-second intervals. Later studies found overestimation of both ten- and thirty-second intervals with the same dose of carbon monoxide (Beard and Grandstaff, 1972). Again, these findings seem to be related in part to boredom and sensory deprivation. Mikulka et al. (1970), O'Donnell et al. (1971), and Stewart et al. (1973) also observed some increase in the estimates of thirty seconds at 9.4% carboxyhaemoglobin. Nevertheless, the findings were discounted for several reasons: (1) differences were only significant for a short period (135 to 150 minutes' exposure), (2) estimates did not worsen with continuing exposure, (3) changes were smaller at 12.5 than at 5 Pa carbon monoxide, and (4) positive findings were only observed under conditions of sensory isolation. When subjects were allowed to walk around and stretch between tests, carbon monoxide induced no changes of time estimation to carboxyhaemoglobin levels as high as 20 to 25% (Stewart et al., 1970, 1975; O'Donnell et al., 1970; Mikulka et al., 1970; Weir et al., 1973).

Learning and Memory

Short-term memory as tested by "digit retention" is unaffected to carboxyhaemoglobin readings of 7.5% (Beard and Grandstaff, 1972; Wright, 1976). In contrast, longer-term memory (memorization of meaningless syllables and recall of digits in reversed order, Fitts and Posner, 1967) deteriorates at 7.2% carboxyhaemoglobin (Bender et al., 1971).

Mental Performance

Schulte (1963) inferred a deterioration of mental performance from slower responses to such tasks as letter and colour recognition, the underlining of plural nouns, the crossing of *t*s, and arithmetic ability. For example, the underlining task required 150 seconds with a carboxyhaemoglobin of 0 to 7% but 210 seconds at 10 to 15%. Likewise, the arithmetic task required 800 seconds at 8% but 1000 seconds at 15% carboxyhaemoglobin. Some subsequent investigators have questioned Schulte's carboxyhaemoglobin values, since initial readings of near 0% were reported even in smokers and 20% readings were seen after exposure to only 10 Pa carbon monoxide (Beard, 1969; O'Donnell et al., 1970). Lewis et al. (1970) also observed a slowing of simple calculations and of comprehension following inhalation of vehicle-polluted air. Bender et al. (1971) found a significant deterioration of letter identification, memorization of meaningless syllables, and ability to reverse digits at 7.2% carboxyhaemoglobin (10 Pa), although no changes were seen in the ability to calculate problems, select shapes, or draw analogies. Likewise, Luria (1977) observed an increased difficulty in the identification of partially masked letters, but only after exposure to 50 Pa of carbon monoxide for one hour. O'Donnell et al. (1971a) found no deterioration in the ability to undertake mental arithmetic, even at 12.7% carboxyhaemoglobin, although it should be stressed that altitude training had given their four subjects some previous experience of hypoxia. Benignus et al. (1977) likewise saw no effect of 12.6% carboxyhaemoglobin upon vigilance in a numeric monitoring task, while Beard and Grandstaff (1972) reported no change in the performance of a visual search and selection task at carboxyhaemoglobin levels of 7 and 8%.

Choice Reaction Time

Choice reaction time requires a decision before a response is initiated. Ramsey (1970) reported a significant slowing of the choice reaction time after ninety minutes' exposure to heavy traffic (3.8 Pa carbon monoxide, estimated carboxyhaemoglobin level 3.5 to 4.0%). In a second study (Ramsey, 1972), the increase of reaction time was rather variable but became significant at 5% carboxyhaemoglobin when data from normal, anaemic, and emphysematous subjects were grouped together. A final report (Ramsey, 1973) showed a lengthening of choice reaction time when healthy young adults were exposed to 65 or 95 Pa carbon monoxide for forty-five minutes (blood carboxyhaemoglobin readings of 7.6 and 11.2%, respectively). In contrast to these positive findings, Fodor and Winneke (1972) observed no change in choice reaction time at a carboxyhaemoglobin level of 5.3%.

Psychomotor Performance

No slowing of arm, wrist, or finger movements has been detected with carboxyhaemoglobin readings of up to 20% (Fodor and Winneke, 1972; Sayers et al., 1929; Weir et al., 1973). With the exception of a report by Ramsay (1972) suggesting an increase of reaction time to a visual stimulus after forty-five minutes' exposure to 30 Pa CO (carboxyhaemoglobin about 4.5%), no changes of simple reaction speed have been seen (Wright et al., 1973; Rummo and Salarnis, 1974; Stewart et al., 1970; Sayers et al., 1929).

Finger dexterity and hand-eye coordination have been evaluated by tapping tests and the ability to manipulate small objects, as in the Crawford collar and pin test and the Purdue pegboard test (where the subject is required to assemble pegs, bushings, and rings). Stewart et al. (1970) reported a significant correlation between impaired performance of the first of these tests and carbon monoxide exposure (2.5, 5.0, and 10.0 Pa for eight hours), but considered this relationship spurious since individual comparisons were not significant and there was no evidence of a correlation when using a related test (Crawford screw test). Johnson et al. (1974) observed a deterioration in the hand-eye coordination of tollbooth workers that was proportional to vehicle exhaust exposure. Fodor and Winneke (1972) reported an unchanged performance of the Purdue pegboad test at carboxyhaemoglobin readings of up to 5.3%, but Bender et al. (1971) noted a significant deterioration of Purdue test scores after three and one half hours' exposure to 10 Pa carbon monoxide (an average carboxyhaemoglobin of 7.2%). Others observed no loss of manual dexterity at carboxyhaemoglobin readings as high as 20% (Stewart et al., 1970; Fodor and Winneke, 1972; Weir et al., 1973). A significant variable in this context may be the difficulty of the required task (Putz et al., 1976, 1979). Putz and his associates linked the hand-eye coordination to a peripheral visual monitoring task; when the combined test was carried out at a high rate of tracking, performance was impaired at 5% carboxyhaemoglobin readings.

Dorcas and Weigand (1929) suggested a deterioration of arm-hand steadiness with five hours' exposure to 40 Pa carbon monoxide. However, other investigators have detected no significant changes of steadiness at carboxyhaemoglobin readings of 13% (Stewart et al., 1970) or 20 to 30% (Sayers et al., 1929). O'Donnell et al. (1971) used a tracking task where a needle had to be kept within prescribed limits while the subject monitored three other dials. Performance showed no change despite carboxyhaemoglobin readings of up to 12.7%; however, critics have objected that the total period of tracking (one minute per trial) was rather short. In a modified version of the test, the task became progressively more difficult with time, and note was kept of the period for which subjects could cope with the task. A transient and statistically insignificant deterioration of scores was then seen with two hours' exposure to 5.0 or 12.5 Pa carbon monoxide (final carboxyhaemoglobin of 3.0 or 6.6%), but performance was unaffected after four and one half hours' exposure to 10 Pa carbon monoxide (Milulka et al., 1970; Hanks, 1970).

No ataxia or loss of postural control was seen despite carboxyhaemoglobin readings of up to 20% (O'Donnell et al., 1971; Weir et al., 1973).

VEHICLE-DRIVING SKILLS

Since the automobile is the prime source of ambient carbon monoxide, it is plainly logical to assess psychomotor skills in the specific context of vehicle operation. Limited information is available from driving test batteries, vehicle simulators, and on-the-road tests. Unfortunately, the results obtained have come under serious criticism due to incomplete publication, weakness of experimental and statistical methodology, and limited relevance of the usual laboratory simulators to the normal driving task.

Forbes et al. (1937) increased blood carboxyhaemoglobin levels by up to 30% and then examined reaction times, coordination, and perception in the context of vehicle driving. By concentrating hard, subjects were still able to perform the various tasks quite well, but they commented that subjective estimates of driving ability indicated some impairment. Normal performance had apparently been maintained at the expense of calling upon "spare" mental capacity (Ashton et al., 1972a, b). Unfortunately the experiment was uncontrolled, and it is arguable that a dose of room air might have produced a similar response. Wright et al. (1973) tested their subjects on a driving simulator. Scores for a number of measures of careful driving habits showed a small deterioration in response to a 3.4% increase of carboxyhaemoglobin; however, most changes were only at the borderline of statistical significance (see Tables XIV and XV). Rummo and Sarlanis (1974) used a simulator that tested responses to the slowing of a lead car; subjects described the procedure as "realistic but boring." After twenty minutes of exposure to 80 Pa carbon monoxide (6 to 8% increase of carboxyhaemoglobin), response times were

Table XIV
NONPARAMETRIC SCORES FOR PSYCHOMOTOR TESTS OF DRIVING SKILLS; RESPONSE OF HEALTHY ADULTS TO 3.4% INCREASE OF BLOOD CARBOXYHAEMOGLOBIN*

Test	Control subjects: Worse	Control subjects: No change or better	Experimental subjects: Worse	Experimental subjects: No change or better	Significance of decrease (1 tailed t test)
Reaction time	2	23	6	19	P<0.005
Night vision	2	23	3	22	n.s.
Glare vision	4	21	4	21	n.s.
Glare recovery	5	20	10	15	P<0.005
Hand steadiness	4	21	3	22	n.s.
Depth perception	3	22	5	20	n.s.
All tests	20	130	31	119	P<0.005

*Based on data of Wright et al., 1973.

Table XV
SCORES DURING OPERATION OF A DRIVING SIMULATOR; RESPONSES OF HEALTHY ADULTS TO 3.4% INCREASE OF BLOOD CARBOXYHAEMOGLOBIN*

Test	Control subjects: Worse	Control subjects: No change or better	Experimental subjects: Worse	Experimental subjects: No change or better	Significance
Emergency actions (braking, accelerator up right swerve)	35	109	26	116	n.s.
Careful driving (brake off, centre steer turn signal gentle brake, brake pump)	12	160	26	147	P<0.005

*Based on data of Wright et al., 1973.

slowed (from 7.8 to 9.6 seconds), and steering wheel reversals were also decreased; nevertheless, changes were again borderline, becoming statistically significant only if data for one smoker were excluded.

On-the-road studies have been conducted by Weir et al. (1973), McFarland (1973), Ray and Rockwell (1970), Rockwell and Ray (1967), and Rockwell and Weir (1975). Although a car was used, the task was somewhat artificial. For reasons of safety, McFarland conducted his tests on a deserted stretch of expressway at speeds of only 48 or 80 km/hr. An attempt was made to reduce spare mental capacity by using a foot-controlled hood to occlude the driver's vision. Ray, Rockwell, and Weir had their subject ride as a passenger in a car towed by a 200-foot length of wire, attempting to detect changes of velocity relative to the lead vehicle. The time needed to report such changes increased from 1.3 seconds in air to 3.3 seconds at 10% carboxyhaemoglobin. Weir et al. (1973) commented on a narrowing of the perceptual field at carboxyhaemoglobin readings of 7 and 12%. There was also a reduction in the number of glances into mirrors prior to overtaking, increased variations in vehicle speed, decreased ability to control following distance, and decreased steering movements at 12% carboxyhaemoglobin. McFarland (1973) commented that subjects required more roadway viewing (as judged by hood openings) when the carboxyhaemoglobin was 17%; however, in their experiments there was no change in the number of steering wheel reversals relative to control data.

There is surprisingly little information on traffic accidents in relation to carbon monoxide exposure (Chovin, 1967; Clayton et al., 1960; Goldsmith et al., 1969; Ury et al., 1972). Attempts to demonstrate a deterioration of specific vehicle driving skills have likewise shown only minor changes (Wright and Shephard 1978b; Wright et al., 1973). Two possible explanations are that standard driving conditions fail to tax the spare capacity of an average driver (Brown and Poulton, 1961), leaving a reserve that can be tapped during carbon monoxide poisoning, and that a modest depression of cerebral function may serve to optimize the arousal of a nervous driver, at least under test conditions (Corcoran, 1965), thus maintaining or even improving performance. Unfortunately, epidemiological research relating traffic accidents to ambient levels of carbon monoxide is complicated by the local nature of pollution, the tremendous excess of alcohol-related accidents, and the difficulty in allowing for effects due to cigarette smoking. Both the mechanics of smoking a cigarette and the resulting heavy carbon monoxide exposure could contribute to an accident, while a further complication is the correlation between cigarette and alcohol consumption. Clayton et al. (1960) found 35% of drivers and pedestrians involved in accidents had a carboxyhaemoglobin reading of over 3%, while Yabroff et al. (1974) also found an increased proportion of carboxyhaemoglobin values greater than 5% among drivers involved in fatal accidents. On the other hand, a study from Los Angeles found that the frequency of vehicular accidents was significantly correlated with the level of oxidant pollution but not with the extent of exposure to carbon monoxide (Ury et al., 1972).

Carbon monoxide poisoning has been implicated in some aircraft accidents. Sources of the gas have included faulty oxygen equipment and cabin contamination by exhaust fumes, smoking, and structural fires (Blackmore, 1974; Howlett and Shephard, 1973; Robinson and Wolfe, 1976). In many of the fires investigated, a large part of the carbon monoxide has apparently accumulated after the crash.

ACUTE AND CHRONIC TOXICITY

Acute Toxicity

Much of the early research upon acute toxicity was concerned with the clinical effects of very severe carbon monoxide poisoning, suicide, and murder attempts (Drinker, 1938; Grut, 1949; Haldane, 1895a; Haldane and Priestley, 1935; Killick, 1940; Silk, 1975; Dinman, 1974). Kerr (1946) pointed out that relatives may on occasion have obscured the evidence of suicide in order to avoid losing insurance payments. More recent reviews of acute toxicity have concentrated on the hazards encountered in firefighting (Teige et al., 1977; Barnard et al., 1976; Burgess et al., 1977; Mastromatteo, 1959a, b), aviation (Howlett and Shephard, 1973), and the home (Wright et al., 1974). Accidents have become all too frequent with conversion of oil to gas furnaces without adequate lining and narrowing of the chimney to accommodate colder and more acid fumes. Severe and fatal exposures are of particular interest as indicating vulnerable body systems; in some instances, functional changes have been brought to light that would have passed unnoticed with inhalation of a lesser dose of carbon monoxide.

Deaths have occurred with prolonged exposure to 0.4% of carbon monoxide. A 1% concentration (1 kPa) is sufficient to induce unconsciousness in fifteen to twenty minutes, and 2% (2 kPa) is fatal very quickly. Since vehicle exhaust contains around 6% carbon monoxide, the operation of a vehicle in an ordinary domestic garage can generage a deadly atmosphere in no more than ten minutes.

Both Haldane (1895a) and Henderson et al. (1921) found a general relationship between symptomatology and the carboxyhaemoglobin reading, suggesting that problems were arising from oxygen want in the tissues. Brief exposure to a very high concentration of carbon monoxide (35 kPa) can produce a frontal headache and a slight sagging of the ST segment of the electrocardiogram (Stewart et al., 1973b). In the more gradual type of poisoning commonly encountered, a carboxyhaemoglobin reading of 20% is associated with a mild sinus-type headache (Stewart et al., 1970), throbbing in the temples and some loss of manual dexterity (Diamond, 1974); "any extra exertion, such as running upstairs, produced a very slight feeling of dizziness and some extra palpitation and hypernoea" (Haldane and Priestley, 1935). At levels of 30 to 40% carboxyhaemoglobin, the headache becomes severe. Nausea, vomiting, and irritability are seen, while the skin assumes a characteristic pinkish ("cherry red") hue, sometimes with areas of erythema. "A slight increase of pulse rate, deeper breathing, and slight palpitations became observable during rest, and running upstairs was followed in about half a minute by dizziness, dimness of vision and abnormally increased breathing and pulse rate. At 40% saturation, these symptoms were more marked, and any exertion had to be made with caution for fear of fainting" (Haldane and Priestley, 1935). At carboxyhaemoglobin levels of 40 to 50%, dimness of vision, chest pain, a euphoric confusion, and uncoordination are likely, while concentrations greater than 50% are liable to cause unconsciousness, a disastrous rise of body temperature (hyperpyrexia), convulsions, and even death. "There was no real discomfort during rest, but the breathing and pulse rate were greatly increased, vision and hearing impaired, and intelligence probably greatly impaired. It was also hardly possible to rise from the chair without assistance. Writing was very bad and spelling uncertain. Movements were very uncertain, and it was difficult to recognize objects distinctly or estimate their distance correctly, so that things a long way off were grasped at in vain. Attempts to go any distance caused failure of the legs and collapse on the floor" (Haldane and Priestley, 1935).

More recent data suggest that 8% of hospital admissions from carbon monoxide poisoning have a fatal outcome (Davidson, 1970). At the highest concentrations (over 50 kPa), death may occur from cardiac arrhythmia before there has been any substantial increase of average blood carboxyhaemoglobin readings. The explanation seems that a slug of virtually oxygen-free blood passes through the coronary circulation, provoking ventricular fibrillation (DeBias et al., 1976). The organs that are most vulnerable to carbon monoxide are the heart and brain. Thomsen and Kjeldsen (1974) described myofibrillar degeneration, with the formation of myelin bodies in the mitochondria, when rabbits were exposed to 10 Pa of carbon monoxide for four hours. Likewise, Longo (1977) reported increased vascularity, oedema, and small "petechial" haemorrhages, with softening of the lenticular nucleus of the midbrain. Skeletal muscle necrosis and renal failure have been observed in comatose patients (Günther, 1921b; Hedinger, 1948; Fowler, 1954; Mautner, 1955; Loughridge et al., 1958; Bessondo and Gray, 1978), although it is unclear whether mechanical pressure had arisen during the coma or local hypoxia had damaged the muscle fibres.

If the patient survives the acute episode, elimination of carbon monoxide occurs with a half-time of 3 to 4 hours. Haldane suggested that in experiments on himself, carboxyhaemoglobin readings dropped from 56% to 20% in three hours (Haldane and Priestley, 1935). Nevertheless, there may be after effects. Haldane complained of headache, nausea, vomiting, and severe depression that he likened to mountain sickness. More recent authors have commented on abnormalities of the electrocardiogram (premature ventricular complexes, atrial fibrillation, and ST changes) and the echocardiogram (impaired movement of the left ventricular wall), these changes persisting for as long as several weeks (Corya et al., 1976). Stewart et al. (1970) complained of an excruciating occipito-frontal headache, maximal 3 to 4 hours after exposure. With severe poisoning, the brain may suffer permanent damage (Baker and Lilly, 1977; Remick and Miles, 1977; Makishima et al., 1977; Barrowcliff and

Knell, 1979; Shillito et al., 1936; Kerr, 1946; Douze, 1977; Smith and Brandon, 1973), although the usual symptoms of such injury (general malaise, headache, personality change, and memory disturbance) are hard to document in the absence of previous information about a given patient. One study noted that 39 of 21,143 patients treated for carbon monoxide poisoning were subsequently confined to mental institutions (Hartmann, 1972), although it is possible that in some of these cases a prior mental disturbance may have led to a suicide attempt.

Chronic Toxicity

Considerable debate continues over the existence of any distinct syndrome of chronic carbon monoxide poisoning (Dinman, 1974). Unfortunately, many of the supposed symptoms are rather vague and nonspecific, while the individuals concerned are often claiming workmen's compensation or some other type of third-party liability insurance. Kerr (1946) writes "it is very noticeable that would-be suicides rarely complain of headache, whilst in the compensation cases it is a marked and predominant symptom."

Some species seem particularly vulnerable to carbon monoxide. Pathological changes have been reported in rabbits after as little as four hours' exposure to 10 Pa of carbon monoxide. Findings have included a hyaline degeneration of the heart muscle, with associated electrocardiographic abnormalities (Ehrich et al., 1944; Kjeldsen et al., 1974; Wanstrup et al., 1969) and degenerative changes or haemorrhage in the cerebral cortex (particularly the limbic system, that region controlling autonomic function), in subcortical grey matter of the corpus striatum (lenticular nuclei), and in the brain stem. Consequences of the brain softening include disturbances of posture, reflexes, and gait. Such changes have been observed in dogs that have been exposed to 10 Pa of carbon monoxide for eleven weeks (Lewey and Drabkin, 1944). In most of the animal experiments, impairments of cardiovascular and central nervous system function have only been seen when exposures have induced sustained blood carboxyhaemoglobin levels of 2 to 30%, and in some instances the abnormalities of the central nervous system could conceivably have been a secondary consequence of cardiac damage. Among positive reports from lesser exposures we may note the following:

1. Pathological changes in the central nervous system with an alteration of chronaxie when rats were exposed to 2.5 Pa of carbon monoxide eight hours per day for ten weeks (Shul'ga, 1961), development of an increased number of spines on interneurons of the caudate nucleus (Norton and Culver, 1977), and abnormalities of peripheral nerve conduction for two to four weeks following near-fatal poisoning (Petagan et al., 1976; Pankow et al., 1975);
2. Brain lesions; thinning, scarring, and degeneration of the myocardium; an abnormal electrocardiogram; and abnormal intraventricular and arterial

pressures when dogs were exposed to 5 to 10 Pa of carbon monoxide for up to six weeks (Preziosi et al., 1970; Lindenberg et al., 1961);

3. A reduction in the spontaneous electrical activity of cerebellar Purkinje cells after their exposure to 50 Pa of carbon monoxide or more (Raybourn et al., 1978);

4. Blood changes, including an increase of fibrinolysis (Kalmez et al., 1977) but also enhancement of the tendency to platelet aggregation with increases of blood viscosity following four to sixty-four hours' exposure to 15 to 50 Pa of carbon monoxide (de Quiros and Hess, 1976; Pfaffinger et al., 1979; Marshall, 1979; Marshall et al., 1978);

5. Cardiac hypertrophy, seen in rats after two weeks of massive exposure (50 Pa of carbon monoxide; Penney et al., 1974);

6. Electrocardiographic abnormalities, such as increased P waves and T-wave inversion, with increased vulnerability to electrically-induced ventricular fibrillation—both acute (two-hour exposure of dogs to 10 Pa of carbon monoxide; Aronow et al., 1978, 1979) and long-term (twenty-four week exposure of monkeys to a similar partial pressure of carbon monoxide; DeBias et al., 1976); and

7. An inhibition of hepatic metabolism of drugs such as nicotine and benzpyrene (50 Pa of carbon monoxide; Altland and Rattner, 1979; Gardner et al., 1978; D.Y. Cooper et al., 1977).

Symptoms have been reported in man with industrial carbon monoxide exposure (drivers of producer gas vehicles), despite carboxyhaemoglobin readings of no more than 5% (Grut, 1949). However, the usual complaints (headache, insomnia, fatigue, irritability, impairment of memory, confusion, dizziness, mood disturbances, cardiac symptoms, digestive disturbances, diarrhoea, hyperhidrosis, thirst, and decrease of libido) are hard to evaluate since the subjects concerned were well aware of their gas exposure. Lindgren (1961) compared exposed groups with controls having similar employment; they also found more frequent headaches among the exposed subjects.

There is general agreement that one long-term consequence of carbon monoxide exposure is an increased red cell count (polycythaemia; Wilson and Schaefer, 1979); however, in contrast with the response to high altitude hypoxia, there is a decrease in the red cell 2:3 diphosphoglycerate, with adverse consequences for the oxygen transporting properties of haemoglobin (M.F. Thomas and Penney, 1977). Other reported long-term effects of carbon monoxide exposure include an increased reticulocyte count (Kjeldsen and Damgaard, 1968b) and serum cholesterol (Kjeldsen and Damgaard, 1968a), a leakage of albumin from the capillaries (Åstrup, 1972; Siggard-Anderson et al., 1968), and platelet thrombosis in the capillaries (Niden and Schultz, 1965) with possible changes of clotting time (Kalmaz et al., 1977; Sairo and El-Attar, 1968), electrolyte excretion (Pauli et al., 1968a, b), and urine production (Kjeldsen and Damgaard, 1968b).

Atherosclerosis

The possibility that chronic carbon monoxide exposure might increase the risk of developing clinical manifestations of atherosclerosis has generated considerable controversy (Schievelbern, 1979; Kannel and Castelli, 1976; Wissler et al., 1976; Cucu, 1978; Turner, 1979; Topping, 1977). One problem is that coronary atherosclerosis is widely prevalent in the middle-aged population, but much of this burden of disease is clinically silent. Moreover, ischemic heart disease is a chronic process, whereas the events susceptible to epidemiological study (infarcts and deaths) are essentially acute manifestations. If carbon monoxide has an effect, it might arise either by increasing the deposition of cholesterol or by precipitation of an acute cardiac emergency.

Åstrup and his associates, in particular, argued strongly for several years that carbon monoxide caused intravascular hypoxia and thus an increased intravascular deposition of cholesterol; such a mechanism would explain the link between smoking and heart disease. Schneiderman and Goldstick (1978) used a computer simulation of the arterial wall to demonstrate that hypoxia of the medial tissue was possible during carbon monoxide exposure. Other investigators have maintained that if cholesterol deposition is influenced at all by carbon monoxide, the effect is only apparent at very high carboxyhaemoglobin readings (20 to 30%) in animals that have been fed diets with a high cholesterol content (R.R. Davies et al., 1976; Armitage et al., 1976; Stender et al., 1977). One recent review suggested carbon monoxide might also inhibit the secretion of very low density lipoproteins or the clearance of chylomicron remnants from the plasma.

Nevertheless, published reports have described atherosclerosis in rabbits (Åstrup, 1969; Åstrup et al., 1967; Birstingl et al., 1970; Kjeldsen et al., 1968, 1972; Thomsen and Kjeldsen, 1974; Kersten, 1976; Wanstrup et al., 1969), squirrel monkeys (Webster et al., 1968), and pigeons (Armitage et al., 1976). Cholesterol accumulated in the vessel walls, and there was an associated splitting of subintimal structures, with the appearance of other changes that could be interpreted as early precursors of atherosclerotic disease—gaps between the cells, infiltration with lipid droplets, and the development of oedema. It was reasoned that these changes reflected local oxygen deprivation, since similar lesions were reported with prolonged hypoxia (Boatman and Carter, 1973) and a normal appearance was preserved if 28% oxygen was given with the carbon monoxide (Kjeldsen et al., 1968). Possible explanations included an increase in the permeability of the vessel walls to lipid (Parving, 1972; Siggaard-Andersen et al., 1968) and an inhibition of the enzyme lipase (Kjeldsen and Damgaard, 1968a). One major difficulty in this type of research is that morphological changes are generally assessed by subjective rating. Åstrup and his colleagues were unable to duplicate many of their original findings when a blind evaluation was carried out (Hugod et al., 1978). In a reevaluation of their earlier work, they admitted that carbon monoxide-mediated endothelial damage was not an important factor in the heart disease of smokers with a normal blood cholesterol.

Theodore et al. (1971) did not observe any significant vascular lesions when a variety of species (monkeys, baboons, beagle dogs, rats, and mice) were exposed to 40 to 50 Pa of carbon monoxide for up to twenty-four weeks. Likewise, Malinow et al. (1976) observed no adverse effects when cynomolgus monkeys were exposed to 50 Pa carbon monoxide for fourteen months, despite use of a semipurified diet containing cholesterol. The quantity of cholesterol in the diet may be a significant variable, since in Japan heavy smokers apparently have a relative immunity to both atherosclerosis and heart attacks (Kagan et al., 1974). Carbon monoxide exposure does not change lipid synthesis in the vessel wall (Sarma et al., 1975), but it does increase the cholesterol uptake of human coronary vessels when these are perfused postmortem (Sarma et al., 1975; Tillmanns et al., 1975). It remains less certain that the usually encountered urban partial pressures of carbon monoxide (0.1 to 1.0 Pa) produce such an effect in a normal person eating a normal diet (Tillmanns et al., 1975). Nevertheless, some authors have suggested that in smokers the likelihood of developing arteriosclerotic heart disease is more closely wedded to carboxyhaemoglobin levels than to cigarette consumption (Wald et al., 1973; Wald, 1975; Wald et al., 1977; Aronow, 1976; but not Heliovaara et al., 1978). Wald et al. (1973) commented that the association persisted after adjustment of their data for age, sex, duration of smoking, serum cholesterol, and reported cigarette consumption. However, it remains arguable that the blood carboxyhaemoglobin is merely serving as an indicator of the extent of smoke inhalation and thus the dose of some other and more toxic constituent of the smoke (Åstrup, 1976).

Several epidemiological studies have attempted to correlate severe occupational exposure to carbon monoxide with the development of atherosclerosis (Rosenman, 1979; Zenz, 1979). Jones and Sinclair (1975) found no increase of arterial disease among blast furnance workers despite carboxyhaemoglobin readings of 4 to 5% in nonsmokers and 8 to 9% in heavy smokers. Dublin and Vane (1947) suggested that firefighters had a higher mortality index for cardiovascular and renal disease than any other occupational group, and Gordon and Roger (1969) commented on disturbances of cardiac rhythm in firemen following carbon monoxide exposure. Mastromatteo (1959b) confirmed there was a significant excess of cardiovascular and renal mortality through a study of death certificates for 271 Canadian firemen. Exercise stress testing has also disclosed a substantial (10%) prevalence of ischaemic electrocardiograms among California firefighters (Barnard et al., 1976). While there is little question that firemen are exposed to carbon monoxide (Levy et al., 1976), it is difficult in most of these studies to discount the influence of other aspects of a firefighting career—long periods of sitting in the fire hall, periodic stressful emergencies, and exposure to a wide variety of toxic fumes. The police are a third vulnerable group. Ayres et al. (1973) noted an ischaemic exercise electrocardiogram in 10% of police patrolling road tunnels; again, the prevalence of abnormal records exceeded expectations, but on the other hand the frequency of definite or possible heart attacks was no greater than predicted from the age and constitution of the sample. Another vulnerable group, in the Shinshu region of Japan, are the workers who make tatami mats (Komatsu, 1955); their work place is often heated by an open charcoal fire, and blood carboxyhaemoglobin readings of 20 to 30% are common. It is thus interesting that atherosclerotic heart disease is five times as common in this population as in residents of other parts of Japan. Other ethnic groups, such as the New Guinea highlanders, live and cook in small and poorly ventilated huts without symptomatic or postmortem evidence of atherosclerosis (Master, 1976).

Attempts to relate the normally modest urban carbon monoxide exposures to cardiovascular deaths (Goldsmith and Landaw, 1968; Goldsmith and Aronow, 1975) have encountered problems due to an uncertain lag period between exposure to the pollutant, prodromal symptoms, and the definitive heart attack. Cohen et al. (1969) found no association between carbon monoxide pressures in the vicinity of Los Angeles hospitals and the number of admissions for myocardial infarction. However, the case fatality rate was related to the atmospheric carbon monoxide level on the day of hospital admission at least in parts of the city that were heavily polluted (weekly average of more than 0.8 Pa carbon monoxide). Nevertheless, detailed examination of their data suggests that the effect described was statistically significant more because of a reduction of deaths in the area of low pollution (case fatality rate dropping from 27 to 20%) than an increase in the area of high pollution (case fatality rate increasing from 26 to 30%). Furthermore, any effect could have arisen not from the pollution per se but from associated interhospital differences in the circumstances of the patients admitted or the standards of medical care. One argument against this last possibility is that the interhospital difference was observed only in the winter months, when pollution levels were high. The correlation could still have arisen from such seasonal factors as an influenza epidemic or even a variation in the number of available hospital beds (since up to 70% of fatal heart attacks and a high proportion of less serious manifestations of ischaemic heart disease are treated outside the hospital system). Hexter and Goldsmith (1971) countered these criticisms by using a sophisticated multiple regression analysis. This related daily deaths to trend and harmonic terms for season, temperature, and oxidant air pollution; the residual effect was small, but it was estimated that if the twenty-four-hour average carbon monoxide pressure increased from 0.73 to 2.02 Pa there would be eleven additional deaths per day in the Los Angeles basin. In contrast to this positive conclusion, Curphey et al. (1965) found no relationship between the cause of death and blood carboxyhaemoglobin levels among 2207 deaths that they examined in the Los Angeles area. Carboxyhaemoglobin readings remain stable after death. However, as Radford (1976) has pointed out, hospital or terminal blood carboxyhaemoglobin levels are not necessarily representative of carbon monoxide exposure either during normal life or in the prodromal period of several days that may precede a fatal heart attack. Kurt et al. (1978) found some increase of cardiorespiratory complaints in the emergency room of the Colorado General Hospital if the twenty-four-hour ambient reading exceeded 0.5 Pa or if the one-hour average exceeded 1.1 Pa; however, blood carboxyhaemoglobins were not determined, and the altitude of the hospital (about 1500 metres) may have increased the susceptibility of the population. A similar type of investigation from the Johns Hopkins Hospital in Baltimore had the advantages of a clearly defined population and well-established diagnoses, but unfortunately pollutant levels were only measured at one sampling station, and average values were quite low (a contrast being drawn between partial pressures of less than 0.4 and more than 0.9 Pa). Although the data showed no clustering of sudden or more protracted deaths from heart attacks during periods when carbon monoxide pressures were increased (Kuller et al., 1975), the postmortem carboxyhaemoglobin readings were greater in arteriosclerotic heart disease deaths than in controls. Although smokers were excluded from the analysis, the authors suggested that their findings were due to a continuing unreported cigarette habit, since the effect disappeared if subjects who described themselves as ex-smokers were also excluded from calculations. Postmortem examinations on 120 of the patients showed no correlation between coronary arterial stenosis and the final blood carboxyhaemoglobin reading.

We may conclude that there is some evidence that regular severe carbon monoxide exposure may increase the risk of developing overt ischaemic heart disease (Goldsmith and Aronow, 1975). On the other hand, the threshold concentration seems to be higher than that normally encountered in the urban environment, unless the individual is sensitized by simultaneous exposure to high altitude.

Chapter 5

VULNERABLE GROUPS

A number of segments of the population are at an above-average risk with respect to carbon monoxide poisoning (*see* Table XVI). In some, the vulnerability has a physiological basis (the foetus, the newborn child, the pregnant woman, and the senior citizen). In other individuals, the problem arises from life-style (the heavy smoker) or disease (for instance, a patient who suffers from exercise-induced angina pectoris). Residence at high altitude may increase susceptibility to tissue oxygen lack. Other air contaminants may also interact with carbon monoxide. Finally, other factors such as exercise-induced hyperventilation and a baseline burden of carboxyhaemoglobin may increase susceptibility.

Some screening of employees is possible in industries where substantial carbon monoxide exposure is likely. However, the sick and the healthy are exposed equally to ambient concentrations of carbon monoxide. The air quality criteria that are accepted for a given community should thus reflect the tolerance of high-risk segments of the population rather than of the healthy young adult who commonly serves as an experimental subject.

THE PREGNANT WOMAN, FOETUS, AND NEWBORN CHILD

The Pregnant Woman

Ethical considerations have precluded the deliberate administration of carbon monoxide to pregnant women. The available evidence is thus limited to

Table XVI
POPULATIONS WITH INCREASED SENSITIVITY TO CARBON MONOXIDE

(a) Hypoxic conditions—foetus, anaemia, chronic obstructive lung disease, ischaemic heart disease, peripheral vascular disease, cerebral vascular disease, high altitude residents, elderly.

(b) Hypermetabolic conditions—increased ventilation due to vigorous physical work, pregnancy, high altitude, thyrotoxicosis, newborn infants, hypoxic conditions.

(c) Increased endogenous production—pregnancy, newborn, haemolytic disease, saturation dives, drugs (phenobarbital, diphenyl hydantoin, methylene chloride).

animal experiments, occasional accidental exposures, and observations made upon women who continued to smoke during pregnancy. In the last group, part of the apparent toxicity to the foetus may in fact be attributable to other harmful constituents of the cigarette smoke plus socioeconomic and genetic correlates of the cigarette habit.

In accidents and attempts at suicide, the pregnant woman herself has a greater susceptibility to carbon monoxide (Smith et al., 1935) due to (1) an increased endogenous production of carbon monoxide, (2) production of carbon monoxide by the developing foetus, (3) an increase of resting respiratory minute volume, and (4) a tendency to anaemia (Hellman and Pritchard, 1971). The fraction of carbon monoxide that is absorbed by the mother penetrates to the foetus unless the exposure is so severe that the mother is herself killed very quickly. Nevertheless, the foetus has some protection, since uptake occurs two to three times more slowly in the foetus than in the mother; maternal readings are not approached for five to six hours, and thirty-six to forty-eight hours are required for complete equilibration (*see* Chapter 3). Unfortunately, this potential advantage is offset by a corresponding slowness of foetal carbon monoxide elimination (Hill et al., 1977). A transient carbon monoxide exposure may yield lower peak carboxyhaemoglobin readings in the foetus than in the mother, but on the other hand the foetus sustains 10 to 25% higher twenty-four-hour average carboxyhaemoglobin readings (Longo, 1970, 1977; E.P. Hill et al., 1977). Engel et al. (1969) found somewhat less carbon monoxide binding by foetal than by maternal haemoglobin, but Longo (1977) observed a 25% higher equilibrium reading in the foetus than in the mother (range 0.6 to 1.6).

The Foetus

Many authors have discussed the harmful effects of carbon monoxide exposure upon foetal development (Wells, 1933; Williams and Smith, 1935; Åstrup et al., 1972; MacMahon et al., 1965; Ginsberg and Myers, 1974; Schwetz et al., 1975, 1979; Abbatiello and Mohrmann, 1979; R. Davie et al., 1972; H. Goldstein, 1971; L.D. Fechter and Annau, 1977; Prigge and Hochrainer, 1977; Hoffman and Campbell, 1978) and perinatal mortality (Longo, 1977; Wood, 1979). Whether the source of carbon monoxide exposure is industrial, urban, or smoking by the mother, it would be helpful to know what problems were caused to early foetal development before pregnancy was confirmed. While foetal hypoxia is most likely in the final trimester of pregnancy, disturbances of foetal growth are more probable if foetal oxygenation is impaired during the first trimester.

Normal patterns of maternal cigarette smoking yield relatively stable foetal carboxyhaemoglobin levels, although there is some decrease at night and in the early morning. Human data has been obtained from umbilical cord blood collected at the time of delivery. In such circumstances, maternal smoking is likely to have been less than during the remainder of pregnancy. Nevertheless, typical readings are 1.1 to 9.2% in children born to smokers, as compared with 0.2 to 3.6% in the offspring of nonsmokers (Heron, 1962).

The harmful effects of a given carboxyhaemoglobin reading are inevitably greater in the foetus because the baseline partial pressure of oxygen in the arterial blood is quite low and there is little possibility of compensating for any further decrease of arterial oxygen content through an increase of cardiac output (Power and Longo, 1975). The main reaction of the foetus to carbon monoxide exposure is a redirection of the circulation towards the brain, heart, and adrenal glands (Longo et al., 1978; Surgeon General, 1979, 1980). The mother who smokes also attempts to compensate for a poor oxygen delivery to her child by developing a larger and a thinner placenta than that found in a nonsmoker (Longo, 1972). However, a large, thin placenta is a mixed blessing for the foetus since it increases the risk of birth complications such as placenta praevia (Christianson, 1979). Moreover, compensation is incomplete, and the smoker's foetus shows a substantial decrease of arterial oxygen pressure relative to that of the nonsmoker (Longo and Hull, 1977). If the foetal blood carboxyhaemoglobin rises to 10%, the arterial oxygen pressure of the developing child drops from a normal value to 20 Torr to a hypoxic 15.5 Torr, while the venous pressure drops from 16 to 12.5 Torr. As in the mother, a shift of the oxygen dissociation curve further hampers release of oxygen from haemoglobin within the tissue capillaries (Longo, 1972). D.P. Davies et al. (1976) calculated that if a mother stopped smoking during pregnancy, a combination of a lowering of blood carboxyhaemoglobin and a normalization of the oxygen dissociation curve would increase the oxygen available to the foetus by 8%.

Harm may result not only from the impairment of oxygen transport within the foetus but also from a derangement of placental metabolism, the latter being observed at a carboxyhaemoglobin level of 7%. The situation of the foetus is further compromised by a threefold increase of endogenous carbon monoxide production in the pregnant woman. The acute effects of carbon monoxide exposure can thus have a disastrous effect upon the unborn child. One study of sheep found that 57% of foetuses sustaining a blood carboxyhaemoglobin of 15% for thirty minutes subsequently died (Longo, 1976; Longo and Hill, 1977). This effect is approximately equal to what would have been observed with a 50% reduction of foetal blood flow (Hill et al., 1977). Ginsberg and Myers (1976) exposed pregnant monkeys to an atmosphere containing 0.1 to 0.3% carbon monoxide (100 to 300 Pa), and found that the oxygen content of foetal blood dropped from 9 to 15 $ml \cdot dl^{-1}$ to less than 2 $ml \cdot dl^{-1}$. Associated changes included a slowing of the heart rate, acidosis, build-up of carbon dioxide, fall of blood pressure, and electrocardiographic evidence of myocardial oxygen lack. Human studies have shown a depression and even a cessation of foetal breathing movements for up to ninety minutes after the mother has been smoking (Gennser et al., 1975; Manning and Feyerabend, 1976).

The chronic exposure of animals to carbon monoxide during pregnancy decreases birthweight and increases perinatal mortality (Åstrup et al., 1972; Fechter and Annau, 1977). There are also long-term effects upon visual evoked potentials (Dyer et al., 1979), voluntary activity (Garvey and Longo, 1978), and the biochemical function of the tissues (Garvey and Longo, 1978; Newby et al., 1978).

The Newborn

There have been few studies of the impact of carbon monoxide upon the newborn infant. Younoszai et al. (1968) noted that immediately following birth, the children of smoking mothers showed a metabolic acidosis. This persisted for up to twelve hours. Åstrup et al. (1972) exposed rabbits to 9 Pa of carbon monoxide for thirty days, inducing a maternal carboxyhaemoglobin of 9 to 10%. In consequence, there was an 11% decrease in the birthweights of offspring, and the newborn mortality rate was more than doubled. Fechter and Annau (1976) exposed rats to 15 Pa of carbon monoxide throughout gestation. The birthweights of the offspring were on average 3% less than normal, behavioural activity prior to weaning was less than anticipated, brain protein levels were abnormally low, and the rate of neonatal weight gain was subnormal. Since exposure to high altitudes also impairs development of the foetus and the young child, there is a risk of additive responses to altitude and carbon-monoxide induced hypoxia (New Mexico State Department of Health, 1975).

Studies on the children of mothers who smoked ten or more cigarettes per day during pregnancy raise the disturbing possibility of more long-term disturbances of cerebral function, secondary to foetal carbon monoxide exposure. The reading attainment of the children concerned is apparently still impaired at the age of seven years (Davie et al., 1972; Goldstein, 1971), and at eleven years of age there remains a three-month retardation of general ability, four-month retardation of reading skills, and five-month retardation of mathematical ability (Butler and Goldstein, 1973). There has also been a report of hyperkinesis (overactivity) associated with minimal brain dysfunction in the infants of mothers who smoke (Denson et al., 1975). If carbon monoxide is indeed responsible for these effects, the critical carboxyhaemoglobin reading has yet to be determined. Various reports have shown maternal values of 4.1 to 6.7% and foetal readings of 3.6 to 7.6% (Coburn, 1977).

Young children remain more vulnerable to carbon monoxide than adults because of (1) a higher resting rate of metabolism, (2) a greater level of habitual activity, and (3) a lower haemoglobin level. A greater tissue demand for oxygen is coupled with a large relative respiratory minute volume (and thus a greater inhalation of carbon monoxide during any peak of pollution). It is debatable whether the child is better or worse equipped than an adult to meet such a challenge; although the cardiac output is smaller (hypokinetic circulation), there is little likelihood of significant peripheral atherosclerosis in a child. One study at a nursery in downtown Chicago noted average carboxyhaemoglobin readings of 6.98% (Behrman et al., 1976).

CIGARETTE SMOKERS AND DISEASE

Cigarette Smoking

Cigarette smoking increases the risk of a given carbon monoxide exposure partly because of the initial body burden of carbon monoxide and partly because it predisposes to a number of diseases in which oxygen transport is impaired (chronic obstructive lung disease, coronary atherosclerosis, angina pectoris, and intermittent claudication).

During the daytime, the blood of a cigarette smoker may contain 5 to 10% carboxyhaemoglobin (Rode et al., 1972). If there is indeed a threshold for effects upon psychomotor performance, the smoker is already at or near this threshold. The margin separating the smoker from a fatal carboxyhaemoglobin of ~ 60% is also substantially reduced. Disturbance of alveolar gas distribution slows the elimination of carbon monoxide following a heavy exposure, and the repeated inhalation of air heavily contaminated by the carbon monoxide of cigarette smoke has the same effect.

Against these factors must be set some measure of acclimatization. As with other types of chronic carbon monoxide exposure, cigarette smoking induces certain adaptive mechanisms, particularly an increase of haemoglobin level.

Hypoxic Diseases

Any disease that impairs the transport of oxygen from the atmosphere to the working tissues compounds the problem created by carbon monoxide exposure. An impairment of pulmonary oxygen uptake or a reduced blood oxygen carrying capacity (conditions such as anaemia, a low cardiac output, or a reduced arterial oxygen pressure) increases the risk of CO poisoning (Bartlett, 1968; Dinman, 1968; Permutt and Farhi, 1969). Some reports from the United States have set the prevalence of atherosclerosis at about 12%, with a total of more than 20% of the population affected by various types of cardiorespiratory disorders; others have set the prevalence of significant stable angina pectoris at no more than 2.1% (Moriyama et al., 1971).

Several of the pathological conditions mentioned, such as anaemia, give rise to hyperventilation and thus a speeding of carbon monoxide uptake (Cropp, 1970). On the other hand, in chronic chest diseases such as emphysema and chronic bronchitis a combination of (1) a reduced pulmonary diffusing capacity, (2) increased erythropoiesis, and (3) a rightward shift of the oxygen dissociation curve all afford some protection against carbon monoxide. In normal individuals, some reduction of arterial oxygen tension may occur during severe carbon monoxide exposure since a low venous oxygen content increases the importance of venous-arterial shunts and a ventilation-perfusion imbalance in the lungs (Brody and Coburn, 1969), while more direct effects arise from the altered shape of the oxygen dissociation curve. Most patients with chronic chest disease show an exaggeration of this phenomenon, with a greater than normal widening of the alveolar-arterial oxygen pressure gradient during CO exposure. Aronow et al. (1977) found that in ten patients with chronic obstructive lung disease, an increase of carboxyhaemoglobin from 1.5% to 4.1% reduced the time to onset of marked exercise-induced shortness of breath (dyspnoea) by 33%. However, it has yet to be established whether the worsening of performance had primarily a respiratory or a cardiac basis. Tissue oxygen lack could lead to lactate accumulation and thus excessive respiratory effort. A widening of the arteriovenous oxygen difference could exaggerate the handicap of intrapulmonary shunts. Finally, a reduced delivery of oxygen to the heart muscle could lead to a loss of myocardial contractility and thus a weakening of pumping by the heart.

In an anaemic individual, not only is there an initial impairment of oxygen transport but exposure to a given dose of carbon monoxide leads to a larger percentage of the available haemoglobin being blocked by carbon monoxide molecules, with a greater potential for extravascular escape of the gas (to myoglobin and to cytochrome P-450). Haemolytic disease of the newborn compounds the normal problems of anaemia with difficulties due to an increased endogenous production of carbon monoxide (Necheles et al., 1976; Oski and Altman, 1962).

Coronary Vascular Disease

Atherosclerotic disease of the coronary vessels increases the vulnerability of the heart to carbon monoxide since the overall oxygen delivery to the myocardium is impaired, while patchy local hypoxia (including sometimes specific lesions of vessels supplying the conducting tissue) leaves the muscle more vulnerable to electrical disturbances (*see* Tables XVII and XVIII). Unfortunately, coronary atherosclerosis has a substantial prevalence among middle-aged workers, and the disease process may be clinically silent for a substantial time (F.H. Epstein, 1973; B. Lown, 1979).

Ayres et al. (1965, 1968, 1969) studied coronary gas pressures during bolus exposure to high carbon monoxide pressures (100 Pa for 8 to 15 minutes, or 5000 Pa for 30 to 45 seconds). The resultant carboxyhaemoglobin levels were no more than 6 to 12%; nevertheless, both normal subjects and patients with

Table XVII
EPIDEMIOLOGICAL EVIDENCE CONCERNING CARBON MONOXIDE EXPOSURE AND DEATHS FROM ISCHAEMIC HEART DISEASE; SUMMARY OF POSITIVE FINDINGS

Criterion	Partial pressure of carbon monoxide and estimated carboxyhaemoglobin	Authors	Comment
Case fatality rate	0.8 Pa (7-day average) (~ 1.4% COHb)	Cohen et al. (1969)	Seen in winters only—mainly *reduced* mortality in area of low pollution
Excess cardio-vascular deaths	2 Pa (24-hour average) (~ 3.5% COHb)	Hexter and Goldsmith (1971)	Allowance for effects of season, temperature, and oxidants
Sudden death, myocardial infarction, total arteriosclerotic heart disease	<0.4 Pa (24-hour average) versus 0.9 Pa (~1.6% COHb)	Kuller et al. (1975)	No effect *except* higher postmortem carboxyhaemoglobin in arteriosclerotic heart deaths (due to subjects describing themselves as ex-smokers)
Cardiorespiratory complaints in emergency room	>0.5 Pa (24-hour average (~0.9 COHb) >1.1 Pa (1-hour average) (~0.6% COHb)	Kurt et al. (1978)	Population sensitized by an altitude of over 1,500 m ?

Table XVIII
REACTIONS TO CARBON MONOXIDE IN PATIENTS WITH ISCHAEMIC HEART DISEASE AND PERIPHERAL VASCULAR DISEASE

Technique	COHb%	Author	Comment
Passive smoking—time to angina	1.8-2.3	Aronow (1978)	Effect could be due to other constituents of smoke or to psychological response to smoke.
Double-blind inhalation of CO—time to angina	2.5-3.0	Aronow and Isbell (1973)	No ECG changes
Double-blind inhalation of CO—time to angina	Δ 2.9	Anderson et al.	No greater effect at 4.5% ΔCOHb. No decrease in exercise time.
Double-blind inhalation of CO—onset and duration of intermittent claudication	2.8	Aronow et al. (1974)	

atherosclerosis showed a widening of the coronary arteriovenous oxygen difference, with an associated reduction of lactate extraction during the passage of blood through the heart. It seems possible that the bolus of carbon monoxide produced some poisoning of myocardial enzymes, and in support of this suggestion reactions were less marked with longer exposure to a lower carbon monoxide pressure, despite an equal ultimate increase of carboxyhaemoglobin (Ayres et al., 1970).

As in animal experiments (Adams et al., 1973; Young and Stone, 1976), normal subjects were able to compensate for the effects of carbon monoxide at least in part by an increase of both cardiac output and coronary perfusion (Adams et al., 1973; Ayres et al., 1968, 1969; Horvath, 1973). Permutt and Farhi (1969) estimated that a 20% increase of coronary blood flow would compensate for a 5% increase of blood carboxyhaemoglobin. In practice, Adams et al. (1973) observed a 13% increase of coronary flow for 4% carboxyhaemoglobin and a 54% increase for 20% carboxyhaemoglobin. In contrast, patients with coronary atherosclerosis are unable to increase coronary blood flow (Hersch, 1964), and probably for this reason they are unable to augment their cardiac outputs to compensate for impairment of oxygen transport.

The difficulty is most obvious in patients with clinical angina pectoris. Such individuals develop a typical hypoxic electrocardiogram during carbon monoxide exposure. Aronow et al. (1974) also found various signs of failing pump function, including an increase of intraventricular end-diastolic pressure, a decrease in the rate of rise of left ventricular pressure (dp/dt), a decrease of stroke volume, and a decrease of cardiac output, in response to 2.05% increase of blood carboxyhaemoglobin.

Aronow and Rokaw (1971) showed that the smoking of lettuce-leaf cigarettes was sufficient to shorten the time to development of angina during subsequent exercise. The carboxyhaemoglobin level increased from 1 to 8%, and it seems likely this change was responsible for the observed effect. The cigarettes were nicotine-free, but other constituents of the smoke or circumstances related to smoking may have hastened the onset of angina. The same type of criticism applies to early freeway tests with anginal patients (Aronow et al., 1972). Exposure to the polluted air was not carried out in a double-blind fashion (Theodore, 1973), and even if psychological reactions to the experiment can be excluded, the individuals concerned undoubtedly inhaled air contaminants other than carbon monoxide. On one day, they were driven along the Los Angeles freeway in a vehicle with open windows, encountering an average carbon monoxide pressure of 5.3 Pa; over a ninety-minute drive, blood carboxyhaemoglobin increased by a surprisingly large 3.96% (to an average of 5.08%; Aronow et al., 1972; Haagen-Smit, 1966; Theodore, 1973). On this occasion, the endurance time on a standard cycle ergometer test dropped from an initial control value of 249 seconds to a figure of 174 seconds. The electrocardiogram also showed earlier ischaemic depression of the ST segment, when the subject had performed less exercise, and at a lower cardiac work rate (as estimated by the systolic blood pressure/heart rate product). On a second occasion, the drive was repeated while breathing compressed air from a gas cylinder. Under these circumstances, the carboxyhaemoglobin reading dropped to 0.65%, and performance of the cycle ergometer test was unchanged relative to the initial control reading.

Aronow and Isbell (1973) later carried out a similar but double-blind experiment in the laboratory, subjects liable to angina pectoris breathing in random order compressed air and compressed air containing carbon monoxide. The concentration of carbon monoxide that was administered was only enough to increase carboxyhaemoglobin readings by an average of 1.65% (total $CO\overline{Hb}$ of 2.5 to 3.0%), but the cycle ergometer endurance time decreased from 224 to 188 seconds. The cardiac work rate at the onset of angina (as estimated from the systolic blood pressure/heart rate product) was also decreased by the carbon monoxide exposure, although no electrocardiogram changes were seen in any of the exercise tests. Passive exposure to cigarette smoke aggravated angina at even lower carboxyhaemoglobin readings (1.8 to 2.3%), but the situation may have been complicated by psychological factors or by physiological reactions to other components of the cigarette smoke (Aronow, 1978).

Anderson et al. (1973) had subjects with stable angina breathe air or carbon monoxide (the latter at a partial pressure of 5 or 10 Pa) for four hours prior to a treadmill walk. The two experimental conditions produced 2.9 and 4.5% increments of blood carboxyhaemoglobin; surprisingly, the time to angina was reduced equally in the two conditions, although the larger dose of carbon monoxide generated a longer-lasting pain. Hypoxic depression of the ST segment of the exercise electrocardiogram was deeper in five of the ten patients after carbon monoxide exposure, but the change was not statistically significant. Neither dose of carbon monoxide influenced the total exercise period tolerated by these subjects.

K.H. Jones and J. Knelson (1980) took the results of both Aronow and Anderson, attempting to fit a continuous function that would allow for differences in both time to angina and initial carboxyhaemoglobin content. They suggested use of the equation

$$\frac{\Delta T}{T_o} = f(\Delta CO\overline{Hb}\%)$$

where T_o was the initial time to angina, ΔT the change in this time, f a constant, and $\Delta CO\overline{Hb}\%$ the change in carboxyhaemoglobin percentage. While the data showed a very wide scatter, a significant linear dose/response regression was demonstrated, at least in the case of Aronow's experiments. Stewart (1974) and Jones and Knelson (1980) have argued from such analyses that there is no threshold dose for the worsening of angina—even a small increase of blood carboxyhaemoglobin produces some adverse effect, and it is thus rather difficult to specify a "safe" ambient level of pollution.

Nevertheless, it has yet to be demonstrated conclusively that exposure to carbon monoxide increases either the frequency or the severity of chest pain. Further, it has yet to be established that carbon monoxide exposure causes other complications or shortens life expectancy in the anginal patient. (Coburn, 1979). Relatively few heart attacks have been reported while smoking (Kuller et al., 1975) or during other types of acute carbon monoxide exposure. On the other hand, it seems generally agreed that minor electrocardiographic abnormalities become more prevalent and/or exacerbated when middle-aged men (either the ostensibly healthy or those with overt ischaemic heart disease) undertake vigorous exercise after exposure to carbon monoxide (E.W. Anderson et al., 1971, 1973; E.W. Anderson and Andelman, 1973; Hackney et al., 1962). This concept is supported by animal studies showing a lowering of threshold for the induction of ventricular fibrillation after prolonged carbon monoxide exposure (10 Pa for twenty-four weeks, with a carboxyhaemoglobin of around 9%; DeBias et al., 1976). Thomsen and Kjeldsen (1974) also described myofibrillar degeneration and myelin body formation in the mitochondria of rabbits exposed to 10 Pa of carbon monoxide for four hours (final carboxyhaemoglobin readings of 8 to 9%). However, not all animal experiments have yielded positive results; studies of the monkey and of the rat did not disclose any harmful effects following a two-year exposure to 10 Pa carbon monoxide (Eckhardt et al., 1972; Stupfel and Bouley, 1970).

Intermittent Claudication

The blood supply to the calf muscles is marginal for exercise in patients with intermittent claudication, and it is logical to anticipate that symptoms would be exacerbated by exposure to carbon monoxide.

Aronow et al. (1974) tested responses to a double-blind experimental exposure; this increased carboxyhaemoglobin readings on experimental days by an average of 1.69% (to a total $\overline{COHb}$ of 2.8%). The maximum tolerated duration of treadmill walking diminished from 174 to 144 seconds under these circumstances.

RESIDENTS AT HIGH ALTITUDE

A substantial proportion of United States residents live at altitudes in excess of 1500 metres, and the same is becoming true in Canada as Calgary and Vancouver expand towards the mountains. Johannesburg is at an elevation of about 2000 metres, and Central and South America have very large cities at even higher altitudes. Several factors increase the risk of carbon monoxide poisoning in such environments, including not only altitude-related hypoxia but also inefficiencies in vehicle operation at reduced atmospheric pressures.

Narrow valleys hemmed in by surrounding mountains may further create thermal inversions, with dangerous "pocketing" of pollution. If the accumulation of carbon monoxide is expressed as a partial pressure (Pa), readings are independent of altitude, but if units of ppm are adopted, a figure of 8.7 ppm at sea level is equivalent to 10.3 ppm at 1500 metres.

The permanent resident of a high-altitude region undergoes some adaptations to the environment, including a compensatory polycythaemia; however, there have been few reports on the carbon monoxide tolerance of high-altitude natives (Master, 1976; Weiser et al., 1979; R.S. Mitchell et al., 1979). Hazards are likely to be much greater for the newcomer (including the middle-aged vacationer). Time is needed to retune both the individual's own cardiovascular system and the compression ratio of his or her car. Altitude-induced hyperventilation favours both an increased intake of carbon monoxide (Forbes et al., 1945) and its rapid subsequent elimination; on the other hand, the decreased oxygen pressure in the blood favours the binding of carboxyhaemoglobin. The rate of elimination of carbon monoxide is thus relatively independent of altitude, at least to 3000 metres (Garodinsky et al., 1970).

The impact of combined altitude and carbon monoxide exposure was examined initially within the context of locomotive operation in the Alps (Haldane and Priestley, 1935) and the flying of light aircraft (where the cockpit or the oxygen system could become contaminated with exhaust fumes). Other situations where oxygen lack has been combined with carbon monoxide accumulation include major fires and the driving of cars through crowded mountain tunnels. Forbes et al. (1945) and Lilienthal et al. (1946) both calculated "carboxyhaemoglobin equivalents" for a combined exposure to hypoxia and CO, while McFarland et al. (1944) and Halperin et al. (1959) developed "physiological altitudes" based upon the total ambient pressure and the blood carboxyhaemoglobin. The altitudes under discussion were fairly high (1500 to 5000 metres); a reduction of visual intensity discrimination was described, the effect of a given increment in carboxyhaemoglobin matching that observed with a corresponding decrement of oxyhaemoglobin; thus an 8 to 10% carboxyhaemoglobin reading reduced altitude tolerance by about 1200 metres. In contrast, Vollmer et al. (1946) found no additional effect from carbon monoxide exposure at simulated altitudes of 3050 and 4570 metres.

Pitts and Pace (1947) based their comparisons upon the heart rates observed at altitudes of 2000 to 3000 metres. They estimated that a 1% increase of carboxyhaemoglobin was equivalent to a 108-metre increase of altitude. At a first reading it might thus be inferred that a modest 1000-metre increase of altitude would induce a disturbance of function equivalent to a 9% increase of baseline carboxyhaemoglobin. However, more recent studies suggest that the hypoxic effect of altitude is nonlinear, with little physiological reaction to altitudes of less than 2500 metres (Buskirk et al., 1967). The main reason for the threshold is the sigmoid shape of the oxygen dissociation curve of haemoglobin (Ernsting and Shephard, 1951), and it thus seems likely that the altitude threshold is modified by CO-induced changes in the shape of the oxygen dissociation curve.

OTHER AIR CONTAMINANTS

Carbon monoxide could interact with other air contaminants either chemically or biologically; while there is some evidence of chemical reactions with peroxides, the data on biological interaction with other urban and industrial pollutants is far less convincing.

Animal experiments have suggested synergism between carbon monoxide and NO_2 (Busey, 1972; Najakima et al., 1972), carbon monoxide and ozone (Murphy, 1964), and carbon monoxide and benzene (Ponsold et al., 1978). However, observations in man have shown no interactions between carbon monoxide (3 to 5 Pa) and peroxyacetylnitrate (Drinkwater et al., 1974), SO_2, or ozone (Hackney et al., 1975). There has been one report of enhanced toxicity to carbon monoxide when exposure occurred in the presence of phenol and dust (Elfinova et al., 1972).

The main theoretical basis for interaction would seem to be an increase of mortality when carbon monoxide poisoning is superimposed upon exposure to an agent causing pulmonary oedema. In such a situation a further impairment of oxygen transport could be lethal. Anticholinesterase drugs and other agents lowering the threshold of cardiac fibrillation could also conceivably exert a cardiac interaction with carbon monoxide.

Lastly, the likelihood of road and rail traffic accidents during periods of dense particulate pollution ("London fogs") may be increased if the vigilance of vehicle operators is impaired by carbon monoxide exposure. Unfortunately, there seem no good figures for such risks.

OTHER FACTORS MODIFYING SUSCEPTIBILITY

Exercise Hyperventilation

The critical blood level of carbon monoxide for the impairment of psychomotor function is unlikely to be reduced if the subject is physically active. Indeed, it is arguable that to the extent that physical activity increases arousal, the threshold for a disturbance of psychomotor performance may rise.

On the other hand, the greater respiratory minute volume of the active individual allows a more rapid attainment of the limiting (equilibrium) carboxyhaemoglobin value, a factor that is of particular importance during exposure to brief peaks of pollution. For example, if the physical activity induces a respiratory minute volume of $30\ l \cdot min^{-1}$, the carboxyhaemoglobin level reached during a one-hour exposure is at least three times as great as that observed under resting conditions.

Drugs

Animal experiments suggest that carbon monoxide exposure attentuates the effect of stimulant drugs (McMillan, 1974) and enhances the action of depressants (Montgomery and Rubin, 1971), including alcohol (Pankow et al., 1974).

A similar interaction with alcohol, tranquillizers, and other depressant drugs might be anticipated in humans. Alcohol is certainly a major factor in both vehicular and industrial accidents, but available evidence does not indicate any additional effect from carbon monoxide exposure, at least to carboxyhaemoglobin levels of 12% (R.D. Stewart et al., 1974; Rockwell and Weir, 1975; Pankow et al., 1974).

Initial Burden

Subjects exposed to a prolonged drive through polluted city streets may reach work with an initial heavy burden of carbon monoxide that leaves them more vulnerable to industrial exposure. The smaller safety margin of the heavy smoker has already been noted. In the opposite sense, industrial exposure may leave an employee with a substantial carboxyhaemoglobin for the homeward journey, augmenting risks that may arise from vehicle exhaust.

Blood donors who smoke pass their burden of 5 to 10% carboxyhaemoglobin to the individual receiving the transfusion (Shields, 1971; Stewart et al., 1973a). The carbon monoxide content of blood persists unchanged during the period of storage (Millar and Gregory, 1972), and this form of CO exposure is a particular hazard for infants undergoing exchange transfusions (Kandall et al., 1973).

Chapter 6

ACCLIMATIZATION TO CARBON MONOXIDE

There has been much confusion in the literature concerning such terms as habituation, acclimation, acclimatization, and adaptation. A brief discussion of terminology will thus precede a consideration of short- and long-term adjustments of the body to carbon monoxide poisoning.

TERMINOLOGY

HABITUATION. When subjects are exposed repeatedly to an adverse environment, the symptoms that are reported usually become less severe. This is probably because the individuals concerned become accustomed to the environment, the observer, and the body reactions, with a diminution of anxiety. It is essentially a form of negative conditioning and has been given the descriptive name *habituation* by Glaser (1966) and others.

Habituation is marked with a subjectively unpleasant stress such as cold. Killick (1936, 1948) described some diminution of symptomatology when she underwent repeated six-hour exposures to 11 to 45 Pa of carbon monoxide over several months. Behavioural factors may also account for some of the observations of Annau (1974) on carbon monoxide and hypoxic hypoxia. Nevertheless, in most reports the symptoms induced by carbon monoxide poisoning have been so sparse as to leave little scope for habituation to occur.

ACCLIMATION. *Acclimation* seems the oldest of the various possible terms, being derived from the French verb *acclimater* (Folk, 1966). One example of early use is provided by Bartley (1878): "acclimation is the spontaneous and natural accommodation to new climatic conditions."

More recently, it has been suggested that the term be distinguished from *acclimatization* by restricting it to experimental situations: "the functional compensation over a period of days to weeks in response to a single environmental factor only, as in controlled experiments" (Hart, 1957; Dill, 1964; Eagan, 1963; Folk, 1966). This rubric would encompass most laboratory observations on the long-term exposure of animals to carbon monoxide.

ACCLIMATIZATION. Historical precedent, extending back to the classical

French physiologist Paul Bert, proposes use of the term *acclimatize* in a similar sense but reflecting a change of total environment (as in many human exposures to carbon monoxide): "the functional compensation over a period of days to weeks to a complex of environmental factors, as in seasonal or climatic factors" (Folk, 1966).

A few ecological investigators have proposed that "the term acclimatization should be reserved for long-term adjustments which involve evolutionary changes over generations as the result of selection by the environment" (Sealander, 1951). However, in the context of human physiology, such a process seems uniformly described as *adaptation*.

ADAPTATION. In some cases a characteristic may favour survival in an adverse environment, and over the course of several generations this characteristic becomes a part of the heritage of a particular strain (Folk, 1966). This type of response is best distinguished as *adaptation*, although some early authors such as Paul Bert wrote of "acclimatization of the race."

SHORT-TERM ACCLIMATIZATION

Since most of the effects of carbon monoxide poisoning are attributable to oxygen lack in key tissues such as the heart and brain, a short-term adjustment to exposure may be possible through an increase of coronary or cerebral blood flow or an increase of peripheral oxygen extraction. The time frame of such an adjustment is measured in minutes or hours rather than in days or weeks, and it is thus not strictly speaking an acclimation or an acclimatization.

The section on psychomotor responses (*see* Chapter 4) discussed several human experiments in which an effect of carbon monoxide was observed during the early stages of a low-level exposure but the response disappeared as the experiment continued. One possible explanation of these findings would be a secondary increase of cerebral flow (although other hypotheses, such as a change of arousal, cannot be excluded).

Animal experiments have given clear evidence for an increase of resting coronary blood flow following carbon monoxide exposure (Ayres et al., 1970; Young and Stone, 1976). However, there is little margin for further increase of coronary blood flow during vigorous exercise, even in healthy individuals, and this type of adjustment is probably denied to the patient with rigid, stenosed coronary vessels as a result of atherosclerotic disease.

LONG-TERM ACCLIMATIZATION

Studies of long-term acclimation and/or acclimatization began in the context of an historical controversy concerning the function of the lungs. Haldane and the physiologists of the Oxford school maintained vigorously that the lung was an active organ, facilitating the transport of oxygen across the alveolar membrane (the "secretion" theory; Haldane and Priestley, 1935).

This hypothesis was disputed with equal vigour by August and Marie Krogh, and apparent experimental support was eventually traced to certain technical problems, including spatial and temporal gradients of alveolar gas composition and difficulties in the accurate determination of pulmonary venous oxygen pressures. Nevertheless, the secretion theory held vogue for many years and spawned the belief that in similar fashion a person could acclimatize to carbon monoxide by developing an active process to facilitate oxygen uptake or to exclude carbon monoxide. More recent discussion has centred on circulatory changes, including an increase of red cell count and an alteration in the shape of the oxygen dissociation curve of haemoglobin.

Haldane and Lorrain Smith (1896) commented that it required about 0.06% of carbon monoxide to produce the 30% saturation of the blood associated with severe symptoms of poisoning. However, at this time they were experimenting upon themselves more or less continuously from day to day. When isolated experiments were undertaken at a later date, they "found the CO much more poisonous, so that 0.04 per cent produced very severe symptoms. In the original experiments, they had apparently become 'acclimatized' without knowing it."

A series of experiments were then conducted upon mice, leading to the apparent conclusion that there was no active inward secretion of oxygen at low percentages of carbon monoxide but that "active secretion began to show itself as soon as the CO-percentage was sufficient to cause symptoms of CO-poisoning" (Haldane and Priestley, 1935). Further experiments on man again showed arterial oxygen pressures above alveolar pressures at a carboxyhaemoglobin reading of 23%, the discrepancy amounting to 8 Torr at rest and as much as 32 to 34 Torr in exercise; in the latter situation, the arterial oxygen pressure was said to exceed the *inspired* value by 15.6 Torr! Esther Killick (1936) was not a supporter of the secretion theory, yet she also commented on a lessening of symptoms, a slower heart rate, and a 30 to 50% reduction of equilibrium carboxyhaemoglobin during successive exposures to the same concentration of carbon monoxide. In vitro experiments ruled out any change in the carbon monoxide-binding properties of haemoglobin, and repeated observations on a closed-circuit system excluded the rapid metabolism of the gas. Killick thus suggested a reversed secretory hypothesis, acclimatization being associated with the development of an active transport of carbon monoxide from the pulmonary capillaries to the alveolar gas. In retrospect, it must be admitted that the available methods of measuring carboxyhaemoglobin and arterial oxygen pressures were very crude during this era, and technical problems were probably responsible for these apparently dramatic findings (Brieger, 1944; Lewey and Drabkin, 1944).

Animal experiments (Campbell, 1929; Nasmith and Graham, 1906; Wilks et al., 1959; Clark and Otis, 1952; Gorbatow and Noro, 1948; Killick, 1937; Zebro et al., 1976) have nevertheless shown evidence of considerable acclimation to carbon monoxide. Over six to fifteen weeks, Killick (1937) brought mice to a state where they would tolerate 200 to 285 Pa of carbon monoxide (60 to 70% carboxyhaemoglobin); controls that had not undergone acclimation showed much more severe symptoms. Surprisingly, Killick (1936) observed no increase of red cell count or haemoglobin concentration with chronic carbon monoxide exposure. However, as in the hypoxia of high-altitude residence, most authors have described some increase of red cell count (polycythaemia; Åstrup et al., 1967; Brieger, 1944; Campbell, 1929; Clark and Otis, 1952; DeBias et al., 1973; Hahn-Pedersen, 1962; Syvertsen and Harris, 1973; Theodore et al., 1971; Zebro et al., 1974), and indeed there have been suggestions of cross-acclimatization between high altitude and carbon monoxide exposure (Clark and Otis, 1952; Wilks et al., 1959). Severe polycythaemia increases the viscosity of the blood and may thus restrict maximum cardiac output. However, more moderate changes increase the oxygen transported for a given cardiac work rate. A typical animal experiment is reported by Vernst et al. (1970). They exposed rhesus monkeys and Beagle dogs to 5 to 50 Pa of carbon monoxide for eleven to twenty-six weeks. The red cell counts, haemoglobin concentrations, and haematocrit readings all started to increase during the first week of exposure and stabilized at a new, elevated level after one month of exposure. The polycythaemia consisted of an increased number of normal red cells (normocytic) with a normal haemoglobin content (normochromic), the response being a linear function of the inspired partial pressure of carbon monoxide. Jones et al. (1971) exposed rats, guinea pigs, monkeys, and dogs to 5.1, 9.5, and 20 Pa of carbon monoxide eight hours per weekday for a total of ninety days; the lowest exposure induced no haematological changes, but an increase of haemoglobin content was seen at the two higher carbon monoxide pressures. An increased tolerance of carbon monoxide can be detected after four to five days, and acclimation continues into the second month; in contrast, some of the tolerance is lost if carbon monoxide is not administered for several days (Gorbatow and Noro, 1948; Zebro et al., 1976). This pattern of response suggests a major role for the increase of red cell count, which develops over one to two months (Syvertsen and Harris, 1973).

The concept of a dose/response threshold is supported by negative findings in monkeys with a carboxyhaemoglobin of 8% (Eckhardt et al., 1972) but positive findings after three weeks at 10 to 15% carboxyhaemoglobin (DeBias et al., 1973).

In humans, data from blood donors (Stewart et al., 1974) and smoking withdrawal clinics (Rode, 1970; Rode et al., 1972) suggest that a modest polycythaemia develops in response to a habitual blood level of 7% carboxyhaemoglobin. During a submarine patrol, the partial pressure of carbon monoxide remained less than 2 Pa; nevertheless, a rise of haemoglobin, haematocrit, and red cell count was observed at thirty-two days, although at fifty-two days all values had returned almost to baseline (Wilson and Schaefer,

1979). The time constant of haematological acclimatization may be relatively slow, since Kjeldsen and Damgaard (1968) were unable to detect any blood changes following eight days of intermittent exposure to 500 Pa of carbon monoxide, despite a mean carboxyhaemoglobin reading of 13%.

There is plainly scope for acclimatization through an alteration of 2:3 diphosphoglycerate concentrations within the red cell. The hypoxia of high altitude and of anaemia leads to an increase of 2:3 DPG; this shifts the oxygen dissociation curve to the left, facilitating unloading of oxygen in the active tissues (Benesch and Benesch, 1967; Palek et al., 1967; Chanutin and Curnish, 1967; Kennedy and Valtis, 1954; Lenfant et al., 1968). However, reactions to the tissue hypoxia of carbon monoxide poisoning are less clear cut. Åstrup et al. (1970) reported a decrease of 2:3 DPG with relatively brief (twenty-four-hour) exposures to high pressures of carbon monoxide (a carboxyhaemoglobin exceeding 20%). Other observers have seen no change (Mulhausen et al., 1968, after eight days at 13% carboxyhaemoglobin; Radford and Kresin, 1973, after two- or twenty-four-hours' exposure to 50 to 100 Pa carbon monoxide) or an increase of 2:3 DPG (Dinman et al., 1970; three hours at 20% carboxyhaemoglobin).

One possible explanation of the conflicting findings is that a moderate increase of carboxyhaemoglobin readings (16 to 20%) increases 2:3 DPG, but that more severe poisoning leads to a decrease; it is arguable that with moderate poisoning the rightward shift of the dissociation curve helps performance but that with more severe exposure a leftward (Haldane) shift is needed to ensure adequate uptake of oxygen in the lungs.

High-altitude exposure leads to other changes in the oxygen transport system, including adjustments of acid/base balance that permit more vigorous ventilation (Kellogg, 1963; Kreuzer, 1967), an increase in the number of both alveoli and pulmonary capillaries (Burri and Weibel, 1971), an increase of pulmonary arterial pressure (Dawson and Grover, 1974), an increase in the capillarity of the tissues (Opitz, 1951; Tenney and Ou, 1970), and an increase in the myoglobin content of the muscle fibres (Vaughan and Pace, 1956; Anthony et al., 1959; Reynafarje, 1962; Hurtado, 1971). There is as yet no information whether carbon monoxide exposure induces similar changes.

POSSIBLE GENETIC ADAPTATIONS

It is unlikely that several generations of humans would undergo exposure to very high and potentially lethal concentrations of carbon monoxide with emergence of a strain peculiarly adapted to such pollution. However, there is some evidence of genetic adaptation to high-altitude hypoxia (Shephard, 1978), and this could conceivably help the individuals concerned if they were exposed to a high concentration of carbon monoxide at a more moderate altitude.

The relative contribution of environment (acclimatization) and genes (adaptation) to the characteristics of the high-altitude native remains controversial. Some advantages, such as a high haemoglobin level (Hurtado, 1964) and an increased cardiac output (Vogel et al., 1974) certainly arise if sea-level residents spend time at low oxygen pressures. Other anatomical characteristics include increased static lung volumes, greater alveolar size, increased pulmonary capillary bed, greater heart volume, and larger bone marrow (Barcroft, 1914; Burri and Weibel, 1971; Bharadway et al., 1973). The carotid bodies are also larger than in sea-level residents (C.W. Edwards, 1971), and there is a reduced sensitivity to the respiratory signals of carbon dioxide accumulation and oxygen lack (Lahiri and Milledge, 1968; H.V. Forster et al., 1969; Dempsey, 1971; Byrne-Quinn et al., 1972; Bisgard et al., 1974; Lahiri et al., 1978). Whether genetic or environmental in origin, such changes would probably afford some protection against carbon monoxide poisoning.

Chapter 7

CURRENT EXPOSURE PATTERNS

In this chapter, an attempt is made to analyse current patterns of exposure to carbon monoxide in the urban, industrial, and domestic settings. The task is far from easy since the prime source of carbon monoxide (the automobile) is highly mobile. Particularly under calm weather conditions, major variations of carbon monoxide concentration occur from moment to moment over a space of a few metres. Inferences about the pattern of exposure can be drawn from an analysis of grab samples or tracings from continuous analysers, but for many purposes it is necessary to use man as an integrating device, estimating the overall daily exposure from blood carboxyhaemoglobin readings (Gilmore and Hanna, 1976).

URBAN EXPOSURE

Sampling Networks and Data Reporting

Most environmental protection agencies operate a limited network of long twin-beam infrared analysers (*see* Chapter 2), although occasional reports of urban conditions are based upon grab samples (see Table XIX; length of stain tubes, samples collected in syringes for analysis in the base laboratory) or continuous electrochemical analysis (the portable battery-operated Ecolyser system).

The infrared detectors are usually arranged to record one-hour, eight-hour, and twenty-four-hour average carbon monoxide concentrations since environmental standards are specified in such terms. The equipment is fairly bulky, sensitive to vibration, and requires a stabilized AC power source. Monitoring is thus generally carried out from a fixed site. In order to indicate the general urban pall rather than local traffic congestion, air should be sampled from an upper window overlooking a quiet street rather than a busy thoroughfare (Colucci and Begemann, 1969; Hamming et al., 1960; Moureau, 1964; Ott and Eliassen, 1973). It remains arguable that the recorded pressures (0.3 to 1.0 Pa carbon monoxide) are not typical of the conditions encountered by some workers in the crowded central core of the metropolis (Brice and Roesler, 1966;

Table XIX
THE INFLUENCE OF SUBJECTIVE RATING OF TRAFFIC DENSITY UPON THE CARBON MONOXIDE PRESSURE FOUND IN GRAB SAMPLES FROM THE SIDEWALK AND THE INSIDE OF A CAR*

Sampling site	Traffic Density and CO pressure (Pa)		
	Heavy	Moderate	Light
Car, Bayview Ave.	3.9	2.4	1.9
Car, Jarvis St.	3.3	1.7	1.4
Sidewalk, Oriole Pkwy.	5.8	1.7	0.6
Sidewalk, Bay St. underpass	6.1	2.0	0.3

*Based on data of Godin et al., 1972.

Godin et al., 1972; McCormick and Xintaras, 1962; Wilkins, 1956; Wright et al., 1975). Microscale readings may thus provide a more valid indicator of risk for specific occupational categories such as taxi drivers, pedestrians, and street workers (see Figure 9).

Individual analysers are relatively costly both to install and to maintain. Sampling sites in the United States increased from 212 in 1973 to 456 in 1977 (Environmental Protection Agency, 1978).

Data interpretation must take account of detector design and calibration techniques (*see* Chapter 2). The data reported is inevitably influenced by the sampling period (hourly, daily, or seasonal). Extended averaging smooths out peaks of pollution that may have some biological significance. Most cities show peaks of pollution coincident with the morning and evening rush hours (Brief et al., 1960; Colucci and Begemann, 1969) plus a late evening "shoulder" of pollution due to increased atmospheric stability (*see* Figure 10; Godin et al., 1972). Superimposed upon the diurnal rhythm are weekly cycles (K.L. Johnson et al., 1968) and longer seasonal variations in carbon monoxide production (increased heating of homes and inefficient operation of cars in the winter months) and seasonal variations in atmospheric stability (the thermal inversions of the fall contrasted with the vigorous winds of the spring equinox). Geographic features such as narrow valleys may also favour the local accumulation of pollutants (Shephard, 1974). Because of these several factors, two sampling sites may show identical mean values but substantially differing peak carbon monoxide pressures.

For simplicity, data are commonly reported as the arithmetic mean. This statistic is heavily influenced by occasional aberrant values (whether due to instrumental failure or an atypical accumulation of pollutant), and the median or the geometric mean is the more appropriate statistic to report (*see* Chapter 2). It is also desirable to ensure that the danger threshold is not surpassed; for this purpose, attention is directed to the second-highest value of the year.

Despite regular maintenance and servicing, most monitoring stations

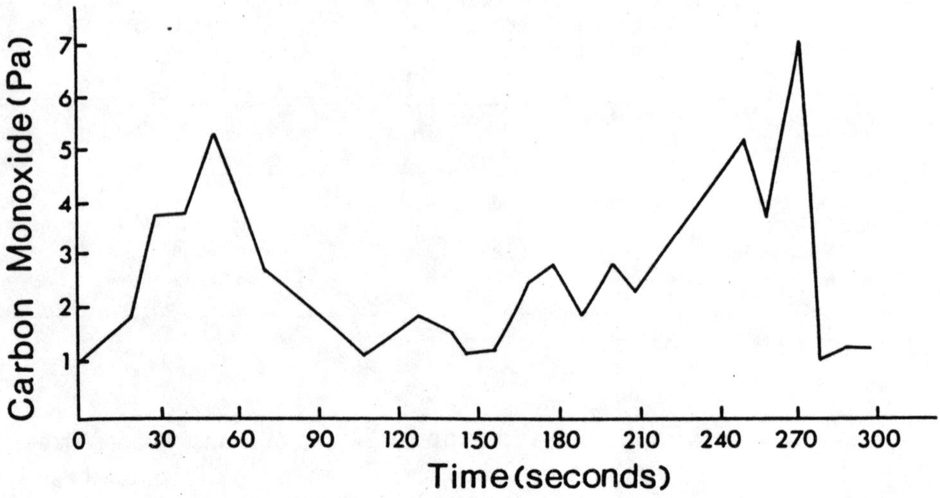

Figure 9. Example of rapidly fluctuating carbon monoxide pressures at a busy intersection in Toronto. Repeated readings from Ecolyser, mean CO pressure 2.6 Pa. Based on data of Wright et al., 1974.

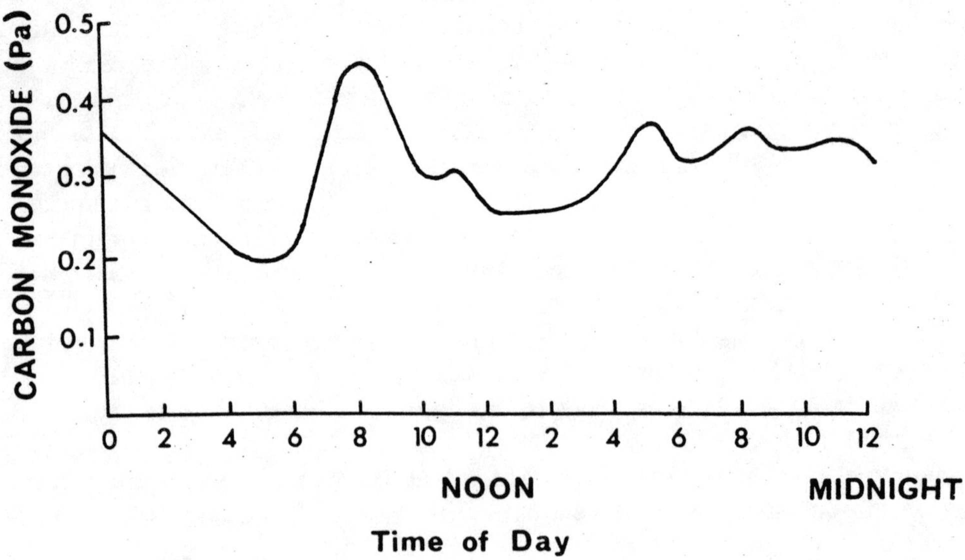

Figure 10. Average hourly carbon monoxide readings obtained from fifth floor of building in downtown Toronto during month of September. Based on data provided by Ontario Department of Health.

develop occasional faults. An attempt can be made to estimate missing data based on the frequency distribution of available carbon monoxide readings. More sophisticated statistical treatment is complicated by autocorrelation of successive observations. For example, a calm evening will inevitably give rise to a sequence of high carbon monoxide readings that persists for several hours, while a weather cycle of high barometric pressure (and thus atmospheric stability) may persist for a week or longer.

Local Urban Readings

The general background of carbon monoxide pollution reported for most large North American cities has been less than 1 Pa; indeed, even long-path analyzers operate for much of their life at close to the limit of their sensitivity. Thus, a survey of 2000 hourly readings from one urban sampling station in Toronto showed only seven readings over 1 Pa (0.35% of data).

Local street-level carbon monoxide concentrations are often higher, ranging from 1 to 5 Pa on busy routes (Bove and Siebenberg, 1970; Brice and Roessler, 1966; Georgii et al., 1967; Godin et al., 1972; Ontario Ministry of the Environment, 1972; Ramsey, 1966; Wilkins, 1956; Wright et al., 1975) with much higher values (5 to 20 Pa) in such sites as overhanging entrances, poorly ventilated tunnels, and underpasses (Ayres et al., 1973; Bithel,1970; Conlee et al., 1967; Lawther et al., 1962; Miranda et al., 1967; Ramsey, 1967; Rispler and Gilbert, 1970; Wilkins 1956; Wright et al., 1975). Chovin and Truffert (1974) made a detailed analysis of conditions in an autoroute tunnel at St. Cloud (Paris); a busy crossroads sometimes halted egress from the tunnel, and despite fans producing twenty air changes per hour carbon monoxide rose to 20 kPa for periods of several hours; moreover, the average reading was 6 kPa from 6 AM to midnight. The data obtained from modern underground garages are usually in the range 0.5 to 4.3 Pa, but occasional values to 70 Pa have been recorded in poorly ventilated corners of less well designed facilities (Wright et al., 1975).

Attenuation is quite rapid on moving away from the source of pollution (*see* Figure 11). Readings are highest a little above the ground, but are already reduced by a factor of 2 to 3 at a height of 1.5 metres (see Table XX), with further decreases on moving up the side of a high rise building (General Electric, 1972). Equally, there may be a three- to fourfold reduction in readings on moving back four metres from the curbside (*see* Table XX). Finally, a two- to eightfold gradient of carbon monoxide pressures occurs on moving from the front of a building facing a busy street to a laneway at its rear; as might be expected, differences were largest in the rush hour but bore little relationship to subjective judgements of wind speed (Godin et al., 1972).

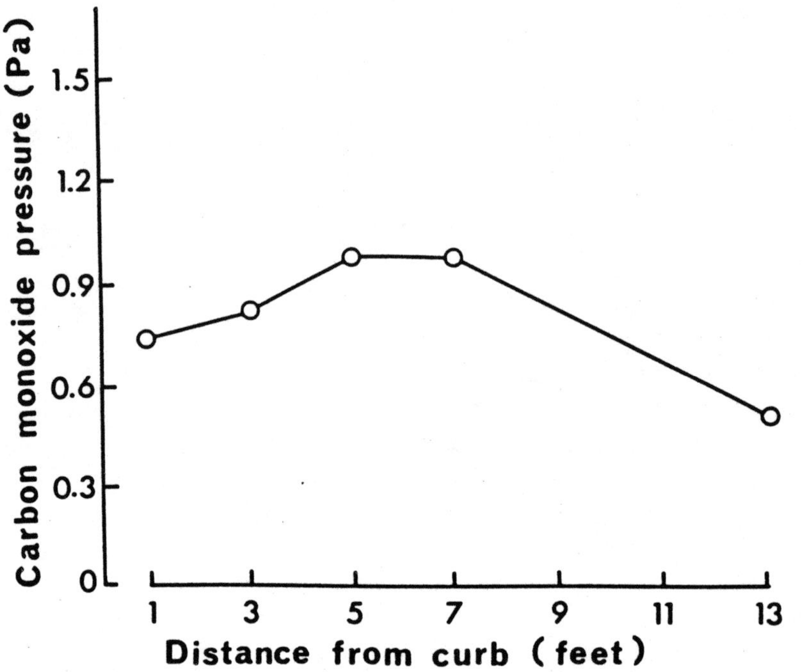

Figure 11. Pressure of carbon monoxide at varying distances from curb of a busy street. Based on data of Wright et al., 1974.

Table XX
INFLUENCE OF DISTANCE FROM CURB, OR DISTANCE FROM GROUND UPON ATTENUATION OF CARBON MONOXIDE PRESSURES*

Distance from curb (metres)	Carbon monoxide pressure (Pa)	Distance from ground (metres)	Carbon monoxide pressure (Pa)
0.3	1.4	0	2.1
0.9	0.9	0.4	3.1
1.5	0.6	0.9	2.1
2.1	0.5	1.2	1.7
4.0	0.4	1.5	1.3

*Based upon data of Wright et al., 1974, 1975.

The protection afforded by the cab of a vehicle varies with the season. In Toronto, vehicles use forced-air heating for some six months of the year, and unfortunately in most designs of car the air intake for such heaters is placed at the front of the hood close to the exhaust pipe of the vehicle ahead; when first entered, the cab may contain no more than the normal urban burden of carbon monoxide, but thirty to sixty seconds of forced ventilation brings the reading close to levels found at the centre of the street. During the summer, when vehicles are driven with the windows open, readings coincide closely with those found on the street (Godin et al., 1972).

Traffic density varies rapidly over the course of a few minutes. It is thus more appropriately assessed by subjective ratings than by official traffic counts. Carbon monoxide readings for both the sidewalk and the cabs of vehicles vary with such subjective ratings (*see* Table XIX). An increase of windspeed helps disperse carbon monoxide: thus readings for one street corner decreased from 1.6 to 0.9 Pa as the windspeed increased from 16 to 32 km/hr (Wright et al., 1974, 1975). However, the clearance of carbon monoxide from narrow streets depends more upon local gusts of wind than the official wind speed as reported by the meteorological department, and subjective ratings are thus useful. Concentrations of carbon monoxide are usually higher on sidewalks to the leeward side of the main vehicle stream. However, if tall offices are built close to the sidewalk, they may function as a windbreak, allowing accumulation of carbon monoxide even when the prevailing wind direction is *towards* the traffic. Furthermore, eddies on the opposite side of the street can lead to a rapid dispersion of carbon monoxide (Nicholson, 1975; Chovin and Truffert, 1974). Despite these complications, significant relationships can be demonstrated between pollutant levels and such measures of atmospheric stability as sunshine hours and the subjective ratings of wind speed (Godin et al., 1972).

General equations for the prediction of urban exposure to carbon monoxide take the following forms:

Car driver $(\log x + 1\ CO) = 0.72 + 0.015(S) + 0.36(U) + 0.077\ (T)$
Car driver $(\log x + 1\ CO) = 0.95 + 0.013(S) - 0.021(W) + 0.082(T)$
Pedestrian $(\log x + 1\ CO) = 0.73 - 0.036(W) + 0.126(T)$
Car driver $(\log x + 1\ CO) = 0.88 + 0.012(S) - 0.035(W) + 0.101(T)$
or Pedestrian

where CO is the carbon monoxide concentration (in ppm), S is the reported hours of sunshine per day, U is the urban carbon monoxide concentration (in ppm), T is the traffic density (rated on a subjective five-point scale), and W is the wind speed (also rated on a subjective five-point scale). The traffic density apparently has more effect upon the exposure of the pedestrian than upon that of the driver (presumably because of the buffering volume of air already inside a vehicle); perhaps because of a specially large cab capacity (N.J. Smith, 1976) and perhaps also because smoking is prohibited on most North American public transit vehicles, bus drivers show less increase of carboxyhaemoglobin than do taxi drivers over a working day. Sunshine provides one simple measure of atmospheric stability; a stable atmosphere holds the carbon monoxide close to the point of emission on the roadway, thus increasing the risk of drivers relative to that of pedestrians (Godin et al., 1972). Nevertheless, the impact of weather conditions is complex, since a fine day also speeds traffic movement whereas wet or stormy weather increases accidents and traffic congestion.

Urban carbon monoxide concentrations show little correlation with levels of some other air pollutants (for example, SO_2, r = −0.05; total oxidants, r = −0.01; oxides of nitrogen, r = −0.01). However, because of their common source, there is a substantial relationship between carbon monoxide and hydrocarbon readings (r = 0.37; Godin et al., 1972).

While some pollutants such as sulphur dioxide emissions produce harmful effects over distances of several hundred miles, carbon monoxide pressures are exceedingly low as little as 10 to 20 km outside of major cities (0.001 to 0.02 Pa) unless there is agricultural burning or a forest fire in the vicinity of the sampling device (Junge, 1963).

Carboxyhaemoglobin Readings

Carboxyhaemoglobin readings provide an index of urban exposure integrated over a period of several hours; nevertheless, most subjects still show appreciable variation in results from day to day. Gettler and Mattice (1933) compared findings for eighteen residents of New York City (1.8 ± 1.2%) and for twelve country dwellers (1.2 ± 1.0%); they commented that values were "nowhere as low as . . . expected." They were unaware of endogenous carbon monoxide production, stating that "the ideal normal individual should have no carbon monoxide in his blood." However, part of the average burden of the country dwellers was undoubtedly attributable to the smokers in the group. Gettler and Mattice (1933) noted the strong association between high carboxyhaemoglobin values and smoking. Hanson and Hastings (1933) confirmed this conclusion, as have numberous subsequent studies (Curphey et al., 1965; Goldsmith and Landaw, 1968; Gothe et al., 1969; Longo, 1970; McIlvaine et al., 1969; Ringold et al., 1962; Rode and Shephard, 1971; Stewart et al., 1973a; Wright and Shephard, 1978a).

P.V. Cole (1974) made a study of seventy-eight blood samples from residents on the Island of Sark in the Channel Islands. In this isolated community no motor cars are permitted. The only petrol vehicle is one invalid carriage operated by the feudal overlord (the Dame); there are also about thirty diesel-fuelled tractors and a few petrol-operated lawn mowers. P.V. Cole (1974) found that ambient carbon monoxide readings were uniformly less than 0.1 Pa. The mean blood carboxyhaemoglobin for the nonsmokers averaged 0.68%, although there was a wide range (from 0.2 to 2.6%), possibly reflecting a few undeclared smokers. The average was a little higher in women (0.70%) than in men (0.61%). Rode and Shephard (1973) had similar findings among nonsmoking Eskimos inhabiting a remote arctic settlement.

Geographical location undoubtedly influences the carboxyhaemoglobin readings for nonsmoking city dwellers (Gettler and Mattice, 1933). High densities of automobile traffic are associated with high carboxyhaemoglobin levels (Rode and Shephard, 1971, 1973; Stewart et al., 1973a; Gothe et al., 1969; Wright et al., 1973). Cole (1974) found an average value of 1.12% COHb for nonsmoking London office workers (most of whom presumably travelled to work by subway). In contrast, nonsmoking meat porters who worked in a large enclosed market (where petrol- and diesel-driven lorries were being unloaded) and who were likely to live in more congested areas of the city had an average reading of 1.51%. The problem of geographical location was highlighted by an extensive spectrophotometric survey of blood donors (Stewart et al., 1973a, 1974). In the 1970 survey, three-quarters of nonsmoking donors from Chicago and Los Angeles had carboxyhaemoglobin saturations of over 1.5%, while in Milwaukee only 26% exceeded this figure. Values were even lower in small towns; thus, in Vermont and New Hampshire only 18% of the nonsmokers tested had carboxyhaemoglobin levels above 1.5% saturation, compared with 45% for the entire sample. Readings also varied with local automobile density; for example, the residents of Governor's Island in New York (average reading for nonsmokers 0.8%) had only half the carbon monoxide burden of those living in adjacent areas of the city. Stewart et al. (1974) further commented that the relative effects of pollution found for differing areas of residence persisted when data for smokers were tabulated in similar fashion; plainly, a polluted residence was hampering elimination of carbon monoxide between cigarettes (*see* Table XXI). Repetition of the blood donor survey in 1974 showed that the proportion of nonsmokers in the Chicago area with more than 1.5% COHb had dropped from 74 to 41%. This was attributed to the success of emission controls (Stewart et al., 1976), although seasonal or sampling effects and systematic errors of analysis* may also be involved. Thus, in March, 1971, Stewart et al. found an average COHb of 1.35% for 673 nonsmoking donors in St. Louis, but Kahn et al. (1974) found an average of only 0.85% for 10,157 nonsmokers from the same city over the period October, 1971 to October, 1972. Curphey et al. (1965) had earlier noted 62% of nonsmokers had carboxyhaemoglobin readings of less than 1%. Kahn et al. (1974), Radford (1976), and Gilli et al. (1978) all concluded that exposure to ambient carbon monoxide added little to the hazards presented by smoking and occupation. Plainly, it is important to resolve these conflicting assessments of normal ambient exposure (Gilmore and Hanna, 1976).

Wright and Shephard (1978b) tested a large sample of visitors to a fall fair in the medium-sized city of London, Ontario (population about 200,000). The surrounding area is pleasant farmland. In this situation, 83% of the nonsmokers tested had a carboxyhaemoglobin of less than 1%, and only 0.7% of the sample had readings over 1.5%. Contrary to the earlier reports of Rode and Shephard

*Radford (1975) drew attention to a 0.6% systematic difference in values reported by the two laboratories when analysing standard blood specimens.

Table XXI
THE INFLUENCE OF AREA OF RESIDENCE AND SMOKING HABITS UPON AVERAGE BLOOD COHb ONE HOUR AFTER LAST CIGARETTE

Area of residence	% Carboxyhaemoglobin and smoking habit		
	nonsmoker	1 pack/day	2 packs/day
Milwaukee	1.2	5.3	4.7
New Hampshire/Vermont	1.4	5.7	5.3
New York City	1.4	4.7	6.3
Washington, D.C.	1.4	5.2	6.6
Los Angeles	2.0	6.0	7.5
Chicago	2.0	6.3	7.7

*Based upon the 1970 data of Stewart et al., 1974.

(1971) and Curphey et al. (1965), they did not find higher readings in men than in women nonsmokers. They suggested that a sex difference of carboxyhaemoglobin was most likely to emerge when the men drove to work through polluted streets while the women remained at home. On the day that the fair was visited, both sexes had a similar drive, and ambient levels of carbon monoxide were low. The data showed little relationship to the reported time spent in an urban versus a rural environment, the annual driving mileage, or the proportion of that mileage spent on urban roads.

Stewart et al. (1973a, 1974) correlated $\overline{COHb}$ readings with a variety of meteorological observations. Data were unrelated to barometric pressure, temperature, or visibility, but wind speed had a slight influence upon the blood carboxyhaemoglobin readings for the community, reducing urban and tending to increase suburban values.

INDUSTRIAL EXPOSURE

Gas-Phase Sampling

The use of gas-phase sampling to specify industrial exposure to carbon monoxide presents many of the problems already discussed in the urban context (*see* Chapter 2).

While some workers occupy a duty station where they are exposed to a consistent partial pressure of carbon monoxide, the majority of occupations at risk involve constant movement about the work place from zones of high pollution to regions where there is relatively little carbon monoxide. Moreover, the varying physical demands of the job lead to corresponding variations of respiratory minute volume, complicating calculations of the rate of equilibration between the blood and carbon monoxide-containing gas pockets in the industrial atmosphere.

Where spatial and temporal variations of pollution are marked, the reported gas pressures should describe the worst situation encountered regularly by an employee over the course of a normal working day. Even this may not provide a total index of danger, since peaks of pollution lasting for ten to sixty minutes may be submerged within an innocuous-looking eight-hour average value.

On occasion it also becomes important to take note of seasonal variations of pollution. Thus the risk to traffic police and other street workers is increased during the winter months, when low temperatures lead to inefficiencies in the operation of motor vehicles. Cold weather may also lead to a closure of doors and windows, increasing the danger to workers in small garages and workshops. At the height of summer, ventilation may again be reduced to economize on air conditioning. In such circumstances, the reported level of pollution should be representative of conditions during the most adverse season.

Where there is suspicion of an industrial problem, a preliminary analysis may be made, using a series of silica (length of stain) detector tubes. Unfortunately, such indicators give only a crude, semiquantitative estimate of pollution, with errors of up to 20% when testing carbon monoxide partial pressures of 5 Pa. More recently, a portable electrochemical detector, the Ecolyser, has also found wide use. Lastly, if there is a well-recognised risk of dangerous conditions, a continuously operating infrared detector may be installed (*see* Chapter 2).

Personal exposure can be monitored by attaching a battery-operated pump to the worker's belt (S.A. Hall, 1973). While the detector can be calibrated by the use of tanks containing known pressures of carbon monoxide, the accuracy of sampling depends also upon a careful check of pumping rates. The pump speed may decrease as the battery voltage drops, and the sampling rate can also diminish if the airline becomes kinked or clogged by dust. Nevertheless, devices of this type have found application in judging the safety of specific groups of employees, as a basis for epidemiological studies of existing or proposed industrial standards, and as a means of assessing the success of control measures.

Carboxyhaemoglobin determinations based upon venous blood, arterial blood, or even alveolar gas concentrations provide the simplest and most reliable method of integrating exposure over an entire working shift (N.J. Smith, 1977). Such methodology is particularly appropriate when there are wide fluctuations of ambient carbon monoxide pressures. The reported data must be convincing to both management and unions, and it is thus important to exclude the possibility that high carboxyhaemoglobin readings have arisen en route to work or as a consequence of smoking. Decisions must be based upon data for employees who are known nonsmokers, and samples must be obtained on arrival at work as well as the end of the shift (*see* Chapter 2).

Gas-Phase Readings

In theory, the gas-phase readings in most industries are held to the permitted industrial threshold limit value (3.5 to 5.0 Pa in different nations; *see* Chapter 8). However, in practice, emergencies, carelessness, and unrecognized hazards cause much higher exposures, in some cases to lethal concentrations. A shortening of shifts and the use of respirators or compressed air lines are recognized expedients in such situations. Many of the reported gas-phase pressures are summarized in Chapter 1.

Blood-Phase Measurements

With a few important exceptions, the carboxyhaemoglobin levels reached by occupationally exposed nonsmokers are lower than those found among heavy smokers who have jobs where there is no occupational risk of carbon monoxide exposure.

Problems are particularly likely if carbon monoxide is a product of the industrial process, as in gas manufacture and smelting operations. Even under normal working conditions, nonsmoking blast furnace personnel develop carboxyhaemoglobin readings of 2.3 to 14.9% (Davies et al., 1965; Farmer and Crittenden, 1932; Jones and Walters, 1962).

Grut (1949) noted many other working processes that produced high concentrations of carbon monoxide, including casting, welding, drying or preheating, boiler cleaning, chimney sweeping, and the transport of hot coke or slag. In some instances gas accumulated due to the nature of the operation, but often there was evidence of unintended leakage, inadequate installation of equipment, or inadequate ventilation.

Stewart et al. (1973b) examined potential hazards in a large sample of nonsmoking blood donors. The lowest readings were seen in students and housewives. Other groups with limited exposure included "occupations associated with mental health, education, library science, religion, art, road paving and entertainment." The highest readings (average 2.5%, range 1.3 to 5.8%) were in taxi drivers. Those working at major airport terminals also had surprisingly high values. Other groups with readings over 2% included occupations involved with vehicles, metal processing, chemical processing, stone and glass processing, printing, welding, electrical assembly and repair, and certain types of graphic artwork.

A second major occupational problem is the inhalation of vehicle exhaust fumes. One early report noted carboxyhaemoglobin readings of 5 to 13% in police officers patrolling the most congested mercantile districts of Philadelphia (Wilson et al., 1926). Unfortunately, smokers were not distinguished in this study. Subsequent investigations, restricted to nonsmoking policemen (De Bruin, 1967; Bothe et al., 1969; Moureau, 1964; Wright and Shephard, 1978a) and drivers of cabs and delivery vans (De Bruin, 1967; Jones et al., 1972), have not found carboxyhaemoglobin levels greater than 5%. Indeed, in the study of Wright and Shephard (1978a) readings averaged only 2.3 ± 0.4%, with an increase of no more than 0.3% over the working shift. It was thus suggested that under modern conditions the average policeman was less exposed to vehicle exhaust than is generally suspected. Vehicle inspectors are a group that has less escape from idling cars. Hofreuter et al. (1962) found that after work the carboxyhaemoglobin values for nonsmokers (3.4%) and smokers (3.8%) were statistically indistinguishable. Likewise, Cohen et al. (1971a) noted that at the end of their shift, the carboxyhaemoglobin of nonsmoking United States border inspectors had increased from the prework level of 1.5% to 4.8%. Corresponding figures for smokers were 3.6 and 6.4%, respectively. Substantial increases of carboxyhaemoglobin have also been observed among other categories of worker encountering vehicles indoors (*see* Chapter 1 and Table XII).

INDOOR ENVIRONMENT

Gas-Phase Sampling

Accumulation of carbon monoxide in the indoor environment is an important practical concern of the environmentalist (Sterling and Kobayashi, 1977). Most of the population spends a large fraction of the day inside various buildings, and indoor life is particularly characteristic of vulnerable groups such as the very young child, the sick, and the elderly.

Current statistics for the United States (DuBois, 1973) suggest that every year 10,000 persons seek medical attention or lose one or more days of normal activity because of excessive exposure to carbon monoxide. Some 1400 individuals die from accidents involving carbon monoxide, and there are a further 2300 suicides; in addition, a large proportion of the 7300 deaths from

Table XXII
FINAL BLOOD CARBOXYHAEMOGLOBIN LEVELS
OF WORKERS ENCOUNTERING VEHICLES OVER WORKING SHIFT

Occupational group	City	Final $\overline{COHb}$ (%)	Author
Policemen	Paris	<4.8	Moureau (1964)
Policemen	Rotterdam	1.2	De Bruin (1967)
	Amsterdam	1.7	
Policemen	Stockholm	1.3	Gothe et al. (1969)
Parking attendants	Dayton, Ohio	7.3	Ramsey (1967)
Vehicle inspectors	Cincinnati	3.4	Hofreuter et al. (1962)
Customs officers	Buffalo	3.6	Conlee et al. (1967)
Bicycle messengers	Amsterdam	1.9	De Bruin (1967)
City drivers	London	2.3	Jones et al. (1973)
Street cleaners	New York	<1.4	Gettler and Mattice (1933)

*Based in part on material collected by Wright and Shephard, 1978.

fires are really attributable to carbon monoxide asphyxiation. Some 70% of carbon monoxide-related deaths occur in the home. Traditionally, such occurrences were most common in the winter, although in more recent years air conditioners have served to pump auto exhaust into homes, motel rooms, and other vehicles.

Morrison (1974) found that automobiles were responsible for 68% of episodes of fatal poisoning, while heating equipment accounted for 26% of deaths. Among heating episodes, 11% were caused by furnaces, 19% by space heaters, and 64% by other types of heater or unidentified equipment. Problems have also arisen on occasion from refrigerators, stoves, and charcoal fires (Master, 1976). In the last two years, difficulties have been reported with the conversion of oil to gas central heating; gas burns with a cooler flame, and a narrowing of the chimney is essential for proper exhaustion of fumes. Likewise, poisoning has sometimes been traced to the use of a wood/coal mixture on a coal furnace that had been nominally converted to gas operation.

With modern heating units, the usual hazards are improper installation of the ventilation system or use of a fuel other than that recommended by the manufacturer. Decorative heaters such as gas "logs" may generate a high concentration of carbon monoxide because the conditions of combustion are adjusted to yield a realistic-looking flame rather than efficient burning of gas. Difficulty is particularly likely if the unit is not pushed sufficiently far into an existing chimney or the heat of the flame is insufficient to counter downdrafts.

Other sources of carbon monoxide in the home include smoking, cooking appliances, and gas lamps (see Chapter 1). The impact of room ventilation upon the accumulation of carbon monoxide from these sources is substantial. Cuddeback et al. (1976) found time-weighted CO pressures of 1.2 and 1.7 Pa for two taverns and traced this difference to respective ventilation rates of six and one to two air changes per hour. Likewise, Harke et al. (1975) found a CO pressure of 3.8 Pa in an unventilated $43 m^3$ room occupied by eight smokers, but in another room of $80 m^3$ with a ventilation of $42.5 m^3$ per occupant-hour the CO pressure was only 0.45 Pa despite the presence of seven smokers. Because domestic gas ovens are not normally ventilated to outside air, they should not be used as a means of heating a room. The use of charcoal barbecues on balconies and in garages is a potential source of carbon monoxide; impingement of the flame on metal and the pyrolysis of food add to the CO burden. Charcoal should not be used for indoor cooking even in a fireplace, since the heat is insufficient to assure adequate exhaustion of toxic fumes. Camper lamps have sometimes generated carbon monoxide pressures of 20 Pa inside tents; the assumption is that the canvas wall will allow permeation of gases, but the porosity of the tent can be reduced by water, ice, or snow, while synthetic fabrics are much less permeable to gases than are traditional canvas tents.

Penetration of carbon monoxide from sources external to a building depends upon the concentration gradient and the tightness with which the building is sealed. High pressures of carbon monoxide develop if a car is operated in a garage (Trompeo et al., 1964; Ramsay, 1967), and the fumes can penetrate an adjacent house or apartment building (Yocom et al., 1971). Some municipalities no longer allow a direct doorway access to a garage for this reason. In the absence of such a local source, indoor concentrations lag outdoor values by one to two hours (Yocom et al., 1971; Godin et al., 1972; Phair et al., 1958). Figures are usually highest if the dwelling is close to a congested arterial route (*see* Chapter 1 and above), and in the suburbs, indoor concentrations can exceed those found out of doors (see Figure 12). However, under conditions of a thermal inversion, the pall of pollution can spread in a more uniform fashion across the city (Godin et al., 1972). If the air intake values for a building are near traffic, the time lag may be reduced or abolished (*see* Figure 12); the ideal arrangement is to place the air intake on the roof, with a slight positive pressure to exclude traffic fumes that might otherwise enter the structure through doors and elevator shafts (*see* Table XXIII).

Radford (1973) commented that the carbon monoxide pressure was higher than 1 Pa in 20% of the East Baltimore homes that he sampled; in 4%, readings exceeded 2 Pa, and in 1% they exceeded 5 Pa. The problem was most common in rented homes and generally seemed attributable to the use of unvented space heaters or stoves for heating; other authors in the United States have also incriminated the space heater (Amiro, 1969). Morrison (1972) noted that 5 to 10% of dilapidated homes had carbon monoxide concentrations that were sufficiently high to cause illness or death. The energy crisis has increased the likelihood of difficulty as the tenants of badly insulated homes cover windows and walls with plastic in an attempt to conserve heat; such action not only favours an accumulation of carbon monoxide but also reduces the efficiency of combustion in heating appliances. Yates (1967) has stressed the need to fit space heaters with regulators that maintain gas pressure and ensure efficient combustion.

Most houses and apartments have a relatively large cubic capacity, and the greatest risk is inevitably found in small dwellings. In Victorian England, the typical victim of a fractured gas pipe or a carelessly used coin-operated gas meter was an old person living in a tightly sealed single room on his or her own. More recently, victims have been numerous in small structures such as houseboats with winter awnings, mobile homes, and camping vehicles. The hazard of incomplete combustion increases with altitude. For example, propane lamps produce 0.6 $ml \cdot min^{-1}$ of carbon monoxide at sea level, but 60 $ml \cdot min^{-1}$ at 3350 metres.

The currently recommended limit for domestic air contamination (DuBois, 1973) is a time-weighted average of 1.5 Pa; this would induce, after several hours of exposure, a carboxyhaemoglobin of 2.6% in the typical adult nonsmoker, the total including the endogenous burden of 0.5% $\overline{COHb}$.

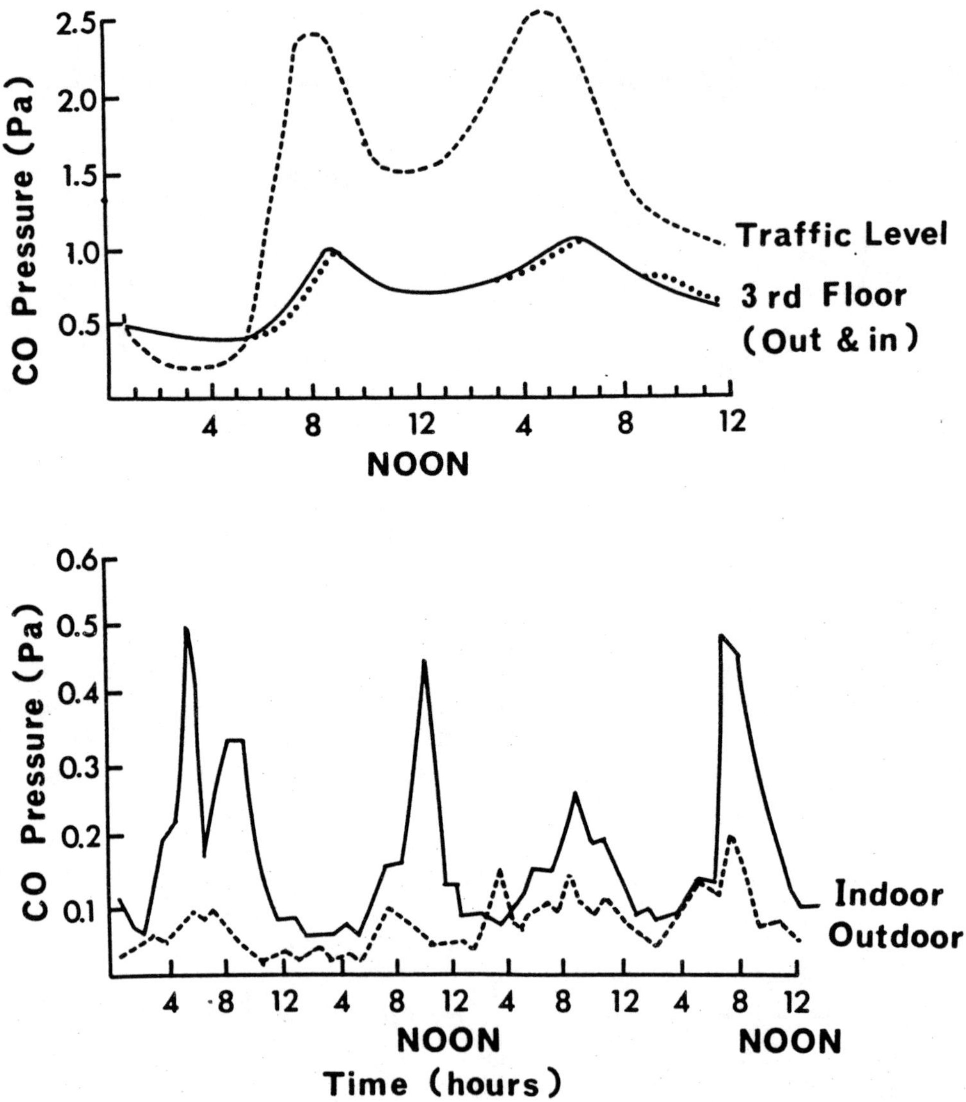

Figures 12a and 12B. Relationship of outdoor to indoor carbon monoxide pressures. Based on data of General Electric, 1972, for urban location (Figure 12a) and Benson et al., 1972, for suburban location (Figure 12b).

Blood-Phase Measurements

The only systematic study of carboxyhaemoglobin levels has related data to cigarette-generated carbon monoxide. Increments of COHb during occupancy of polluted rooms have been used as an overall index of exposure to cigarette

Table XXIII
THE INFLUENCE OF AIR INTAKE ON INDOOR CARBON MONOXIDE CONCENTRATIONS; FLOOR IS SHOWN IN PARENTHESIS

Ventilating system	CO pressure by floor (Pa)				
No air conditioning	1.2 (1)	1.2 (2)			
	2.8 (1)	2.5 (2)	2.0 (4)	1.9 (6)	
Air intake on each floor	5.0 (1)	3.4 (2)	2.2 (3)	2.4 (4)	2.2 (5)
Air intake on roof	2.0 (1)	1.7 (4)	1.8 (7)		
	0.8 (1)	0.7 (7)	0.7 (15)		

*Based on data of N.J. Smith, 1976 for city of Calgary, 5:30 to 6:00 PM, December 21, 1973.

smoke. While the determinations are relatively simple, there is the difficulty that the carbon monoxide persists in the air for longer than particulate contaminants (which have a greater importance for the chronic toxicity of the smoke).

Srch (1967) sat two nonsmokers in a stationary vehicle with two chain smokers. The carbon monoxide pressure within the car reached 6 Pa within twenty minutes and then climbed more slowly to a final value of 8 Pa at one hour. The final blood carboxyhaemoglobin of the nonsmokers was a substantial 5%. Harke (1970) arranged for 105 cigarettes to be smoked in an unventilated room over a period of two hours. The carboxyhaemoglobin of nonsmokers rose from 0.9 to 2.1% with this exposure. When a similar number of cigarettes was consumed but with adequate ventilation, the CO pressure was only 0.5 Pa; however, even this was sufficient to increase carboxyhaemoglobin levels from 1.1 to 1.6%. Russell (1973, 1975) brought the carbon monoxide pressure within a small room to 3.8 Pa through a combination of actual smoking and the smouldering of cigarettes in ash trays. Seventy-eight minutes of exposure to this environment raised the COHb of twelve nonsmokers from 1.6 to 2.6%, equivalent to the response anticipated with the actual smoking of nine cigarettes. Aronow (1978) had three volunteers smoke fifteen cigarettes over the course of two hours. In the absence of ventilation of the room, the carboxyhaemoglobin of nonsmokers increased from 1.30 to 2.28%, and even given the relatively brisk ventilation of eleven air changes per hour the body burden was augmented from 1.26 to 1.77% COHb. Seppanen (1977b) found the COHb of nonsmokers increased from 1.6 to 2.2% over ninety minutes' occupancy of a crowded room where smokers were consuming one cigarette every fifteen minutes. He estimated that over an eight-hour period, each nonsmoking occupant of the room would be exposed to the equivalent of five cigarettes.

One negative report is from Anderson and Dalhamn (1973). They made observations under real life conditions of ventilation and smoking and reported no change of COHb when five nonsmokers shared a room with seven smokers. Although the latter were said to be heavy smokers, ambient carbon monoxide pressures were held to the surprisingly low figure of 0.045 Pa. More recently, Szadkowski et al. (1976) examined the venous blood of office workers who were nonsmokers; in those who shared rooms with smokers, the blood $\overline{\text{COHb}}$ decreased from 0.82 to 0.63% between 8 AM and 3 PM, comparable values for those in smoke-free rooms being 0.92 and 0.72%.

The observations discussed above seem to indicate that an accumulation of cigarette smoke can increase the blood carboxyhaemoglobin readings of nonsmokers, but only if initial values are low and ventilation of the room is poor (Wakeham, 1977a, b). Much depends upon the recent exposure history of the individual. If a subject has encountered 5 Pa of CO while driving through heavy traffic, some of this gas will be eliminated from the body in a room that is moderately contaminated by cigarette smoke (2.5 Pa). The main penalty of the passive exposure to cigarette smoke would be a less rapid and less complete elimination of carbon monoxide than would be possible if the individual had been in clean air.

Basis of Prevention

Maintenance and inspection are important to a preventive programme. Fuel jets of heating appliances quickly deteriorate and require careful adjustment for complete combustion of fuel. Modern instrumentation is also needed to detect small leaks in flues. Legislation on the venting of gas appliances is only effective if backed by regular inspection; ducting is too often blocked deliberately (to conserve heat) or accidentally (by such items as falling masonry, birds' nests, leaves, ice, or snow). Unfortunately, when legislation is introduced it often does not refer to appliances that are already installed; commonly, these are most in need of inspection. The current American National Standards Institute requirement is that a room heater produce no carbon monoxide; this requirement is regarded as met if (1) the carbon monoxide content of the air-free products of combustion is less than 20 kPa when the heater is operating in a room with normal oxygen supply and (2) a carbon monoxide pressure of less than 50 Pa is produced in a closed 1000 ft^3 (28.3 m^3) room during combustion of sufficient gas to reduce the oxygen concentration to 15.1% ($\sim$ 15 kPa). There is a need to develop bylaws to enforce equally efficient and safe heating of rental homes for the urban poor. Given minimum temperature legislation for such accommodation, the sale of unvented space heaters could reasonably be prohibited. Both regulation and publicity could also forestall emergencies associated with use of fuels other than those for which a heating unit is designed. More than 50% of new homes in the United States are now of modular, prefabricated construction, and there is a particular need for uniform federal design requirements so that substandard units cannot be shipped across state boundaries. Equipment that is liable to malfunction at high altitudes should be clearly marked, and its use in permanent installations should be prohibited. Lastly, the public should be warned of dangers arising from paint strippers and the reduction in ventilation of homes.

Chapter 8

LEGAL AND PRACTICAL IMPLICATIONS

This final chapter addresses the legal and practical implications of the data that have been previously presented. Air quality criteria are suggested for urban, industrial, and domestic settings. The possibility of legislating against pollution of enclosed structures by cigarette smoke is discussed. Possibilities for the reduction of personal exposure are reviewed in the context of urban architecture, vehicle design, and control of industrial processes. Lastly, the possibility of successful litigation is examined in relation to various types of poisoning.

AIR QUALITY CRITERIA

Jurisdictional Comment

As noted in Chapter 1, the British government had evinced an interest in the accumulation of carbon monoxide in mines and railway tunnels in the late nineteenth century. However, in other countries, progress was less rapid. In Canada the first ambient air quality criteria were only proposed in 1930, following an investigation of several industrial pollutants (Katz and Drowley, 1966). One problem in federal governments such as Canada's is a division of responsibility. Currently, the federal government recommends air quality criteria for the overall protection of public health and welfare, under the provisions of the "Clean Air Act" (Lucas, 1972). Nevertheless, with a few specific exceptions, responsibility for regulation and enforcement is deputed to provincial and municipal authorities (Katz and Drowley, 1966). For example, in the province of Ontario, the "Air Pollution Control Act" of 1958 gives the provincial Minister of the Environment power to investigate air pollution problems, recommend methods of monitoring air contamination, and assist municipalities in the legal technicalities of drafting effective air pollution control bylaws (*Ontario Gazette*, 1974).

In practice, it has been left to municipalities to regulate emissions from most types of fuel-burning equipment and incinerators, while industrial emissions have fallen under direct provincial jurisdiction. Such a sharing of regulatory authority seems an inevitable consequence of a federal constitution. Nevertheless, it has undesirable practical consequences, as follows:

1. Standards of air quality recommended by the federal government are not necessarily accepted or implemented by lower levels of government. If an industry is confronted by standards that vary from one jurisdiction to another, there is a potential for honest confusion concerning both legal and ethically desirable standards. Moreover, unscrupulous companies with high emission rates may play one province against another when deciding where new factories should be built and where outdated factories should be closed.
2. Pollution is no respector of provincial or even of national boundaries. A nation such as Canada must face the fact that a major fraction of its total pollutant burden is emitted by its more powerful southerly neighbour. Equally, on the local scene a province or a municipality may need to control contamination originating in an adjacent jurisdiction.
3. The regional concentration of authority leads to a focus upon local problems to the possible neglect of global consequences of ever-increasing air contamination; thus, a high stack may transfer pollutants across a provincial or even a national border without reducing total accumulation in the global ecosystem.

Societal Objectives and Air Quality

The choice of air quality objectives depends not only upon scientific facts and their interpretation but also upon the values of a society. How much is a community prepared to sacrifice in terms of economic progress and a supposed standard of material well-being to conserve the lakes of a neighbour, its own clean air, or indeed the comfort of a minority of its population affected by chronic disease (Biersteker, 1972)? The slogan of the Victorians as they raped the Yorkshire dales was "where there's muck, there's money." In the case of carbon monoxide, the social argument assumes new poignancy as western nations face the economic pressures of diminishing fossil fuel reserves and the postindustrial society. In particular, there is a sharp conflict between the need for cheap transportation in sprawling cities designed for the automobile and the poor gasoline mileage afforded by many engines equipped with efficient pollution control devices (Heuss et al., 1971; National Academy of Sciences, 1974). When setting standards for a community, should one seek the complete safety of all citizens or merely the "reasonable protection of public health" (Shy, 1973)?

Various models of the interrelationship between air pollution and human health have been developed to emphasize the consequences of adopting a particular air quality standard (D.O. Anderson, 1967; Colucci et al., 1973; Goldsmith, 1962; Goldstein, 1975; Hatch, 1962; Phair and Sterling, 1969). Since carbon monoxide is colourless and odourless, aesthetic judgements on the value of a clear sky and the fresh scent of the countryside are irrelevant. The gas is also sufficiently inert chemically that the damaging of historic buildings is not a concern. High doses have some effects upon bacteria and leaf formation, but the limiting factor upon which standards must be based is undoubtedly the influence of carbon monoxide upon human performance and/or health.

In proposing solutions, it is important to ensure that other facets of health do not become endangered. Thus, there was some fear that the use of catalytic converters and of diesel fuel in place of gasoline would control the carbon monoxide pollution of large cities at the expense of generating lethal clouds of sulphur dioxide and sulphur trioxide (D.W. Clark and MacMahon, 1981). For reasons which are still debated, this particular fear seems to have been groundless (Maugh, 1977).

Levels of Health Effect

It would be much easier to set an air quality standard if the response to a pollutant such as carbon monoxide showed a clear threshold below which no adverse effects occurred. Clark and MacMahon (1981) suggest that all United States Congressional air quality legislation is based upon such a concept. Some authors such as Horvath and his associates (1975) have maintained that one early sign of carbon monoxide poisoning (a reduction in the individual's maximum oxygen intake) follows this type of pattern. No change of oxygen transport is observed below a carboxyhaemoglobin of 4%, and the deterioration of function is linearly related to the accumulation of carbon monoxide above a carboxyhaemoglobin of 6%. Under resting conditions, it is certainly conceivable that compensatory mechanisms such as a widening of arteriovenous oxygen difference could maintain the normality of most body functions, offering the possibility of a small threshold dose of carbon monoxide. During maximum effort, such compensation is rather less likely, although again there is a high-altitude threshold (around 2000 metres) below which hypoxic hypoxia has little effect. If, indeed, there is no threshold, then it becomes arguable that *any* accumulation of the pollutant is undesirable. One must then establish an arbitrary acceptable level of risk, possibly by setting a monetary value upon human health or life (Duck, 1973; Goldberg, 1972). If the concept of a threshold is accepted, it must equally be recognised that this will differ according to the physiological function that is being examined. Value judgements are still needed to decide *which* functions are of significance to health. Analyses to date (*see* Chapter 4) suggest that the most vulnerable parts of the body are the heart, the brain, and exercising muscle. Disturbances of cardiac function (leading to ventricular fibrillation) and cerebral function (leading to accidents) probably have greater significance than an impairment of exercise performance caused by effects upon oxygen transport and the muscle cell. Comparative analysis suggests five possible levels of human response to carbon monoxide:

Level A invokes some physiological response, but the effect is transient and easily overlooked. In essence, the exposure lies within the potential range for homeostatic adjustment; detection depends on the test sensitivity and the number of observations that are made.

Russian environmental toxicologists are apparently influenced strongly by this Pavlovian concept (Glass, 1975; Ryazanov, 1962). They argue that a pollutant must have no direct or indirect harmful or unpleasant effect upon man. There must be no impairment of working capacity, mood, or physical well-being. Any need for "adaptation" (*see* Chapter 6—actually, a physiological adjustment or short-term acclimatization) is regarded as deleterious, and correspondingly stringent air quality standards are stipulated. The degree of enforcement of such regulations remains unknown. Casual visitors to the Soviet Union form an impression of very low vehicle densities in wide city streets. However, there also appears to be much industrial pollution from low, poorly designed, and poorly sited chimneys.

Level B gives rise to definite and persistent physiological effects, although the observed disturbances (such as a deterioration of time discrimination) may in themselves have no immediate significance for health (Heuss et al., 1971). In a free enterprise society such as Canada (where cars, including the Russian-built *Lada*, are in plentiful supply), there is much discussion of the extent to which an individual's physiological reserve can be exploited in the interests of economic progress (Stern, 1965). Some physiologists have argued that if physiological effects can be demonstrated in a healthy young adult, pollution is already too severe, since no margin has been allowed to accommodate the more sensitive members of the population.

Level C is regarded as the warning level. There is now some physiological evidence of strain. The body's efforts at compensation are becoming noticeable to the subject. Although exposures are tolerable, there may be unpleasant symptoms, including (in the case of carbon monoxide) headache and shortness of breath.

Levels D and E are indicative of an emergency situation, with a risk of pathological changes and death, respectively. In the context of carbon monoxide exposure, an increase of hospital admissions for cardiac disease may be anticipated, and the exposed population will show an excess mortality (Biersteker, 1976). The mortality associated with acute bouts of poisoning is identified fairly readily, but it is more difficult to assess slight trends of chronic mortality. One analysis estimated that existing carbon monoxide levels in the United States generated some 4000 deaths per annum (National Academy of Sciences, 1974).

Industrial versus Community Standards

Levels of health and human responsiveness to any pollutant are distributed asymetrically about a median line (Goldsmith, 1962). Thus, even low-level exposures to a given pollutant can have an adverse impact on the health status of a few sensitive members of a population. On this basis, Goldsmith (1962) reasoned that community standards should reflect an earliest effect for sensitive individuals rather than a no-effect level for the healthy. The United States Congress has accepted this view, in essence legislating protection even for patients with advanced cardiorespiratory disease as long as they are not in a hospital or other specially controlled environments (Clark and MacMahon, 1981).

On the other hand, somewhat greater exposures may be tolerable in industry, since exposure is for only a part of the day, medical examinations can screen out many of the more sensitive subjects, and work in a polluted environment is (for many employees) a largely voluntary decision. Nevertheless, Hatch (1973) maintained that industrial air quality standards should also be set below the limit of physiological compensation, in order to allow a margin of safety for (1) unintentional more severe exposures, (2) individual impairment as a result of age, recent illness, or exposure to other environmental hazards, and (3) possible cumulative effects. Beard and Wertheim (1967) likewise argued that the ability to perform a given task is a valid criterion; certainly, in some skilled operations a deterioration of psychomotor function could lead to costly and dangerous accidents.

The precise ratio of occupational to ambient standards is another hotly debated philosophic issue. With some pollutants, a level that is safe for industry has been established, and a substantial additional margin of protection has then been proposed for the urban environment. For example, a ratio of 30 : 1 has been suggested for NO_2 (Morrow, 1975; Guidotti, 1978). Taking the commonly applied industrial standard for carbon monoxide (a partial pressure of 3 to 5 Pa), this would imply that the eight-hour urban level should be reduced to 0.10 to 0.17 Pa. Such a range would be almost impossible to attain in the average city, and indeed would leave little margin over the natural levels found in the open countryside. A more realistic approach for carbon monoxide is to decide upon an unsatisfactory carboxyhaemoglobin (on either physiological or pathological grounds), to estimate the ambient pattern of exposure likely to cause the average healthy person to reach this limiting value, and then to apply some safety factor to allow for interindividual differences in both exposure and sensitivity to carbon monoxide. Unfortunately, commonly applied safety margins are impracticable for carbon monoxide. Morrow (1975) has argued the parallel between contamination of a common foodstuff and the intake of contaminated urban air. Although concentrations may vary quite widely with time, the city dweller has little respite from urban air contaminants. The United States Food and Drug Administration has proposed that in such a situation, there should be a 100-fold margin between a dose *not* producing any effect in two species of animals and a dose considered as safe. Applying this rule to carbon monoxide, the threshold pressure would be about 2.5 Pa and the safe pressure a totally unrealistic 0.025 Pa. Critics of this approach have suggested that while a 100-fold safety margin is desirable to protect the general population from death, carcinogenesis, and teratogenesis, it seems unnecessarily large for protection against some of the commonly reported manifestations of carbon monoxide poisoning, such as impairment of time judgement.

URBAN STANDARDS

General Considerations

Canadian federal recommendations for the urban environment (Ouellet, 1978) specify three possible control levels: the maximum desirable, maximum acceptable, and maximum tolerable levels (Environment Canada, 1973). It is arguable that a triple standard is confusing and a single requirement might be preferable. The Ontario government in fact specifies required levels corresponding to the maximum acceptable objective. The United States (Environmental Protection Agency, 1971) also specifies a maximum permissible level, while noting higher levels of contamination corresponding to a warning level, an emergency, and a health hazard.

The *maximum desirable level* of the Canadian federal authorities corresponds to Level I as specified by the World Health Organization (WHO, 1964), a concentration of pollutant that can be detected but induces no physiological changes or other adverse effects upon human health. In many respects it is comparable with Level A above. The *maximum acceptable level* is a level of pollution where full compensation is possible (Hatch, 1962). Compensation may, nevertheless, subject the body to some strain; it is thus analogous to level B above. The *maximum tolerable level* indicates the limit of physiological compensation (Hatch, 1962); a further increase of pollution would lead to pathological change. Depending upon the pollutant under consideration, adverse effects may include a deterioration of visibility, damage to vegetation, and an impairment of sensory function (Goldsmith, 1962). Discomfort may be sufficient to cause the exposed individuals to change their place of residence.

Excessive Carboxyhaemoglobin Levels

Normal Adults

The prime criterion of an excessive carboxyhaemoglobin level in the normal, healthy adult seems to be the deterioration of psychomotor functioning that some authors have apparently demonstrated when the carboxyhaemoglobin of the nonsmoker increases to 3 to 4%. Exercise performance is also measurably impaired at about the same level of carbon monoxide exposure.

Detection of a threshold exposure for effects upon the central nervous system is hampered by available test procedures, which show prolonged learning curves and a wide variability of scores. Furthermore, the brain has a substantial reserve capacity that can be called upon in such emergencies as an impairment of local oxygen delivery. If the subject is aroused (anxious concerning the experiment, taxed by the test procedure, or animated by friends), the reserve capacity may be brought into action, but if the subject is bored or in sensory isolation (exposure to carbon monoxide in an audiometric booth or when driving alone), the reserve is not invoked, and the harmful effects of carbon monoxide are more readily demonstrated.

Under resting conditions, some compensation for the central nervous effects of carbon monoxide poisoning is possible through use of the cerebral reserve, increase of arousal, and augmentation of cerebral blood flow. The effects of carbon monoxide upon psychomotor performance may thus be quite transient, and indeed long-lasting disturbances of function are unlikely until these several potential mechanisms of adaptation have been overloaded. The earliest detectable disturbance of function (*see* Chapter 4) seems to be a deterioration of vigilance (failure to detect a difference in light brightness, tone loudness, or signal length), with parallel difficulties in time estimation. This pattern of disturbance mirrors the changes of cerebral function observed during extensive wartime studies of altitude-induced hypoxia. The critical blood level of carboxyhaemoglobin unfortunately remains unclear; various reports suggest threshold saturations ranging from 1.8 to 7.6% (*see* Table XIII). In a number of studies where a high threshold has been described, the reason for the finding has been a failure to test responses at lower pressures of carbon monoxide or the use of an insufficient sample of subjects relative to the variability of the test methodology. On the other hand, the results of Beard and his colleagues at very low carbon monoxide pressures have also been regarded as anomalous since the arousal of their subjects was artificially reduced by infrequent repetition of a boring task within the confines of a small audiometric booth. It remains arguable that such a task is similar to the demands of solo highway driving, a situation where significant carbon monoxide exposure is particularly likely to occur. The probability of detecting impaired vigilance depends greatly upon test circumstances; nevertheless, there seem reasonable grounds for the inference that healthy young adults will show some response at carboxyhaemoglobin levels as low as 3 to 4%. Although the immediate, measured changes are physiological, more serious health consequences could arise from the resultant deterioration of vehicle-driving skills. Unfortunately, evidence concerning driving ability is to date equivocal. The techniques of operating a laboratory driving simulator differ substantially from the skills needed in normal operation of a car, and it is perhaps not surprising that the deterioration of performance is only at the borderline of statistical significance with a substantial (3.4%) increase of blood carboxyhaemoglobin. Likewise, road tests have been on empty highways, without risk, and under such circumstances large doses of carbon monoxide have been necessary to induce perceptual narrowing (7% carboxyhaemoglobin) and decreased use of the rearview mirror (12% carboxyhaemoglobin). Nevertheless, the margin of cerebral reserve is probably diminished by smaller doses of CO, a factor that could be critical when responding to an emergency situation; McFarland (1973) gave a graphic demonstration of one aspect of this diminished reserve in that subjects fitted with a facial visor demanded more roadway viewing time after carbon monoxide exposure.

There is as yet no convincing research linking carbon monoxide exposure with the frequency of vehicle accidents; there remains scope for a large-scale epidemiological study that relates blood carboxyhaemoglobin to traffic accidents after making independent allowance for smoking habits and alcohol consumption.

Deterioration of effort tolerance has also been demonstrated statistically at blood carboxyhaemoglobin levels of 4% and above, although it is still debated whether there is a distinct threshold, as occurs with high-altitude hypoxia; conceivably, a significant deterioration of performance might be demonstrated at a lesser carboxyhaemoglobin given a sufficient sample size. Nevertheless, the presently demonstrated threshold of 4% reinforces the urban air quality criterion set by the vigilance data. Again, the response is physiological rather than pathological and thus defines a maximum acceptable range of pollution. The loss of oxygen transport has practical consequences for the endurance athlete and the person performing a long shift of heavy physical work, but in an ordinary sedentary individual it merely reduces an unused exercise reserve.

Sensitive Population Groups

The group with the highest carboxyhaemoglobin levels is cigarette smokers. However, such individuals should not be considered specifically when setting urban air quality criteria, partly because their baseline hazard is self-imposed (Berlin and Smeets, 1973), partly because the smoker is unlikely to absorb further carbon monoxide at normally encountered ambient pressures, and partly because adaptation seems to occur with chronic smoking. Likewise, workers exposed to high levels of carbon monoxide are unable to absorb further amounts of this gas from the general urban atmosphere. Ambient contamination affects them only insofar as it restricts the elimination of occupationally absorbed carbon monoxide.

The foetus is more sensitive to carbon monoxide than the adult, and some of the phenomena associated with maternal smoking (increased abortion rates, augmented perinatal mortality, and impaired intellectual development of the young child) are plainly pathological rather than physiological disturbances. It is likely that carbon monoxide-induced hypoxia is to blame, although contributory factors could include other constituents of the cigarette smoke, socioeconomic status, and poor health habits in women who are continuing smokers. In animals exposed to known concentrations of pure carbon monoxide, the critical dose for a pathological change is greater than that associated with psychomotor disturbances in the healthy adult (impaired placental function at 7%, foetal death at 15% carboxyhaemoglobin).

A substantial segment of the population suffers from various forms of hypoxic disease. However, such disease increases sensitivity to carbon monoxide only if it limits oxygen transport to those tissues most vulnerable to carbon monoxide poisoning (the brain, heart, and skeletal muscle). The most obvious candidate pathology is cerebrovascular disease, but surprisingly no tests of carbon monoxide sensitivity have been carried out on subjects affected by this disorder. In contrast, there has been considerable investigation of interactions between coronary atherosclerosis and urban carbon monoxide exposure. Possible sequelae of carbon monoxide poisoning include both physiological and pathological responses (a shortening of the time to exercise-induced angina, a lowering of the threshold for cardiac dysrhythmias, and an increase in the likelihood of cardiac death). Passive smoking experiments have suggested a hastening of angina at 1.8% carboxyhaemoglobin, although in such circumstances other smoke constituents or a psychological reaction to the cigarette smoke could have been responsible (Aronow, 1978). A similar shortening of the time to angina has been observed in double-blind carbon monoxide inhalation experiments at 2.5 to 3.0% carboxyhaemoglobin (Aronow and Isbell, 1973), although a little surprisingly symptoms were not accompanied by electrocardiographic evidence of hypoxia such as ST segmental depression, and there was no other evidence of harm to the subjects. Intermittent claudication also appears earlier if the blood carboxyhaemoglobin is increased to 2.8% (Aronow et al., 1974). All of these reports originated in one laboratory, but the shortened time to angina was confirmed by Anderson et al. (1973) in a more severe exposure (subjects exposed to 5 Pa carbon monoxide for four hours, with 2.9% increase of carboxyhaemoglobin). Animal experiments suggest that substantially higher readings (8 to 9% carboxyhaemoglobin) are needed to induce ventricular dysrhythmias and pathological changes in the myocardium. A 2.5 to 3.0% carboxyhaemoglobin must nevertheless be regarded as evidence of at least a Level C exposure (warning level, subjectively unpleasant for patients with coronary and peripheral vascular disease). The proportion of the urban population that is at risk due to cardiac disease is still hotly debated. While 20% of the population have some degree of myocardial ischaemia, it is probable that less than 2% have sufficient vascular disease to be adversely affected by carbon monoxide exposure.

There have been suggestions that a similar carbon monoxide exposure (sufficient to induce a carboxyhaemoglobin of 2 to 3%) increases the likelihood of cardiac deaths, presumably by precipitating ventricular dysrhythmias (*see* Table XVII). If such conclusions are confirmed, this would set an ultimate limit to allowable urban contamination. In some reports, Levels D and E (increased hospitalization and death) have been reached with 1.4% carboxyhaemoglobin at sea level (Cohen et al., 1969) and only 0.9% at an altitude of 1500 metres (Kurt et al., 1978). Nevertheless, the available data must be viewed with some skepticism. There is no guarantee that patients developing cardiorespiratory complaints or dying have not been exposed to substantially higher concentrations of carbon monoxide than those indicated by a limited number of urban sampling stations; furthermore, the regional and seasonal differences of cardiac pathology attributed to carbon monoxide could have arisen from some other unmeasured variable such as a seasonal epidemic or the accumulation of some other air contaminant in the more polluted sectors of the cities studied. It is particularly unsatisfactory that in the most dramatic report (from Los Angeles), the observed difference of cardiac fatality rates was due more to an improved mortality experience in an area of low pollution than to a worsening of fatality rates in highly polluted parts of the city. The only positive finding in Baltimore (where partial pressures of carbon monoxide ranged from 0 to 0.4 Pa) was a high blood carboxyhaemoglobin in nonsmokers dying of arteriosclerotic heart disease (Kuller et al., 1975); even this finding was thought to be due to an incorrect reporting of smoking habits, since the effect disappeared when subjects who described themselves as exsmokers were eliminated from the sample.

Other conditions that increase sensitivity to carbon monoxide poisoning have been listed in Table XVI. Information on thresholds in such groups remains sketchy or nonexistent. All of these groups are presumably more sensitive to carbon monoxide-induced psychomotor disturbances than are normal sedentary young adults. On the other hand, there is as yet no evidence that they are more sensitive than patients with coronary or peripheral vascular disease in terms of the development of pathological problems during episodes of intense pollution.

PHYSICAL ACTIVITY is unlikely to reduce the threshold carboxyhaemoglobin for the development of impaired psychomotor function; indeed, the exercise-induced increase of arousal may raise the threshold. However, the associated increase of respiratory minute volume allows a more rapid attainment of the limiting carboxyhaemoglobin during transient peaks of pollution. If activity is sufficient to induce a ventilation of $30\ l \cdot min^{-1}$, the critical level of 3 to 4% carboxyhaemoglobin will be reached with fifteen minutes of exposure to 20 Pa of carbon monoxide, a sixty-minute exposure to 5 Pa, or an eight-hour exposure to 2 Pa.

HIGH-ALTITUDE SOJOURN is a significant concern, since the effects of altitude and carbon monoxide are necessarily additive (although a simple linear addition of effects is not necessarily appropriate). The situation of permanent residents at high altitudes is further complicated by a probable acclimatization to high-altitude hypoxia. It is difficult to accept the view of Pitts and Pace (1947) that an ascent of 109 metres is equivalent to a 1% increase of blood carboxyhaemoglobin, except at relatively great altitudes. Temporary residents of Mexico City (altitude about 2200 m) show a loss in oxygen transport of no more than 7 to 8% (Goddard, 1967; Shephard, 1967), and on this basis 1% carboxyhaemoglobin is equivalent to 300 metres altitude; the nonsmoking residents of Calgary (altitude 1050 m) thus start their working day with the equivalent of 3.5% carboxyhaemoglobin (0.5% endogenous production and 3% altitude effect). Nevertheless, in the absence of acclimatization, the altitude resident is probably vulnerable to psychomotor disturbances with very small additional doses of carbon monoxide.

The foetus is particularly handicapped at altitude, as suggested by statistics on birth weights and neonatal mortality.

Ischaemic heart disease patients are likewise extremely vulnerable if they happen to visit high-altitude regions. Kurt et al. (1975) reported that a blood carboxyhaemoglobin level of 0.6 to 0.9% was sufficient to increase the hospital admissions of patients with cardiorespiratory disease at an altitude of 1500 metres.

ENDOGENOUS PRODUCTION of carboxyhaemoglobin is rarely sufficient to reach the threshold for disturbed psychomotor function. Even in pregnancy (where CO formation is increased threefold), the baseline is unlikely to exceed 1.5% COHb in a nonsmoker. Nevertheless, the total quantity of carbon monoxide needed to reach a critical blood carboxyhaemoglobin reading is reduced by as much as a third, increasing susceptibility to brief ambient exposures.

HYPOXIC DISEASES such as anaemia and chronic obstructive lung disease probably cause some increase in sensitivity to carbon monoxide by lowering the maximum oxygen intake. The susceptibility to carbon monoxide transients is further increased by hyperventilation, and in haemolytic anaemia there is also an increased endogenous formation of carbon monoxide.

ISCHAEMIC HEART DISEASE, if severe, is necessarily costly to treat, and has a poor prognosis; moreover, many of those individuals with clinical manifestations are in intensive care units and are thus excluded from consideration when setting air quality criteria. The most pessimistic calculations (Hexter and Goldsmith, 1971) suggest that a twenty-four hour average carbon monoxide pressure of 2 Pa (3.5% carboxyhaemoglobin) would cause eleven excess cardiac deaths per day in the Los Angles area. If this extreme pollution persisted throughout the year, it would thus generate about 0.7 deaths per 1000 people per year. Certainly the risk is substantially smaller than that arising from traffic accidents in a major conurbation, both in terms of frequency of occurrences and years lost of "useful life." Nevertheless, it is arguable that if an episode of carbon monoxide accumulation causes eleven excess deaths in a major city, pollution is too high.

Atmospheric Concentrations

The relative merits of the various equations linking carboxyhaemoglobin and ambient concentrations of carbon monoxide have been discussed in Chapter 3. The equation of Coburn et al. (1965) is generally preferred over its rivals because of the range of physiological variables taken into consideration, although it has the disadvantage that an iteration of calculations is needed to allow for the interdependence of carboxyhaemoglobin and oxyhaemoglobin. Table X indicates the relation between ambient carbon monoxide pressures and COHb readings.

Certain factors that influence the choice of safety margin have been discussed above. The universality of exposure to urban air is a possible argument in favour of a substantial margin, as is the fact that many of the tests of carbon monoxide responsiveness have been conducted on healthy young adults rather than upon more sensitive members of the population. On the other hand, only a small segment of the total population is continuously exposed to high local concentrations of carbon monoxide (as at a busy intersection), and the half-time of carbon monoxide retention within the body is relatively short (three to four hours). Finally, there are questions of economics and practicality. Heavy capital expenditures on mass transit would be needed to achieve more than marginal changes in urban carbon monoxide concentrations, while the potential safety margin is further circumscribed by naturally occurring sources of pollution and by endogenous production.

Given the fairly clear evidence of disturbed function at 3% carboxyhaemoglobin and endogenous carbon monoxide production equivalent to a carboxyhaemoglobin reading of 0.5%, the safety margin cannot usefully exceed a factor of 6. Even this margin would be impossible to attain, except by a nonsmoker living in a rural area with no local carbon monoxide contamination of his or her dwelling. A reasonable compromise seems to limit the blood carboxyhaemoglobin to 1.5%. Except at high altitudes, this would ensure a twofold safety margin with respect to the time to angina and an increase of cardiac fatality rate for patients with ischaemic heart disease, while allowing a rather similar margin of protection against a disturbance of vigilance or effort tolerance in healthy young adults. The corresponding carbon monoxide pressures for a person undertaking heavy work would be less than 5 Pa for fifteen minutes, less than 1.5 Pa for one hour, and less than 0.9 Pa for eight hours (*see* Table X).

It would seem reasonably feasible to set both an eight-hour limit of 0.9 Pa and a one-hour limit of 1.5 Pa (*see* Tables XXIV and XXV). The general urban

Table XXIV
CURRENTLY RECOMMENDED AIR QUALITY CRITERIA

Time Period	Authority						
	WHO (1979)	Canada (Ouellet, 1978)		U.S.A. (Env. Prot. Agency, 1971)	Russia (Krotov, 1975)	Japan (Env. Agency 1975)	W. Germany (Fed. Min. of Interior, 1974)
		Acceptable	Desirable				
15 minutes or less	10 Pa						
30 minutes	5*						
1 hour	2.5*	3.0 Pa*	1.3 Pa*	3.5 Pa†			2.6 Pa††
8 hours	1.3*	1.3*	0.5*	0.87†	0.09 Pa	2.0 Pa*	0.9*
24 hours	1.0*						

*Time-weighted average.
†Maximum not to be exceeded more than once per year.
††95% of cumulative frequency distribution.

Table XXV
PERCENTAGE OF DAYS IN 1973 WHEN 24-HOUR URBAN AIR CRITERION OF ONTARIO (0.8 Pa) WAS EXCEEDED*

City	Approximate Metropolitan population	Percentage of days with excess CO
Edmonton	400,000	0
Calgary	400,000	4
Ottawa/Hull	350,000	0
London (Ont.)	150,000	0
Sudbury	100,000	0
Toronto	2,000,000	10
Cornwall	20,000	0
Sarnia	60,000	0
Montreal	2,500,000	2
Hamilton	400,000	0
Windsor	200,000	2

*Based in part on data of Economic Council, 1974.

background concentration of carbon monoxide rarely exceeds such levels in most major North American cities. In Toronto, for example, hourly readings are usually less than 0.5 Pa, and observations greater than 1.0 Pa are a rarity (Wright, 1976). A twofold safety margin may be somewhat hazardous for such individuals as taxi drivers and street workers, who can encounter five to ten times general urban concentrations of carbon monoxide for up to eight hours per day. The appropriate response to this situation probably lies in a correction of the local accumulation of vehicle exhaust (improved ventilation of tunnels, setback of property from the sidewalk, and elimination of traffic bottlenecks) and/or a selection of healthy workers for such occupations, rather than any attempt at the imposition of harsh and unrealistic restrictions upon overall urban carbon monoxide levels. More evaluation of susceptibility is necessary at high altitudes, but on present evidence it would appear that the acceptable ceiling of carbon monoxide pressures should be halved at altitudes in excess of 100 metres.

Currently Proposed Urban Standards

Currently proposed urban standards are based upon the general considerations outlined above (*see* Table XXIV) with varying additional safety factors. For example, the United States eight-hour maximum permitted average of 8.7 ppm (0.87 Pa) is designed to hold the carboxyhaemoglobin below 1.5 to 2.0%, thus providing a twofold safety margin for most normal adults at sea level.

The Canadian federal government has currently recommended limits based on time-weighted average concentrations of carbon monoxide as follows:

Averaging period	Maximum desirable			Maximum acceptable			Maximum tolerable		
	ppm	$mg{\cdot}m^{-3}$	Pa	ppm	$mg{\cdot}m^{-3}$	Pa	ppm	$mg{\cdot}m^{-3}$	Pa
1 hour	13	15	1.3	30	35	3.0	—	—	—
8 hours	5	6	0.5	13	15	1.3	17	20	1.7

In Ontario, the required levels are 30 ppm (3.0 Pa) for one hour, 13 ppm (1.3 Pa) for eight hours, and 8 ppm (0.8 Pa) for twenty-four hours.

The United States standards have a differing statistical basis, being based upon the second-highest reading of the year, with a ceiling of 35 ppm (3.5 Pa) for one hour and 8.7 ppm (0.87 Pa) for eight hours. Superimposed upon this is the warning level (3.0 Pa for eight hours), the emergency level (4.0 Pa for eight hours), and the health hazard level (5.0 Pa for eight hours). The requirements of other jurisdictions are summarized in Table XXIV.

Environmental control is based mainly upon one-hour and eight-hour averages, since the body can equate relatively brief exposure to high concentrations of carbon monoxide with more sustained exposure to lower levels of pollution. However, to the extent that there is some difference of carbon monoxide distribution between a bolus and a steady-state exposure, there may be a need for at least spot-checks of ambient levels averaged over short periods. The relative merits of the United States standards (which can be exceeded only once per year) and average concentrations (as used in other jurisdictions) have been discussed in Chapter 2.

The safety margin of the proposed CO standard is admittedly quite slender, and some experts have pressed for more exacting limits. They have argued that the currently proposed requirements offer insufficient protection against the speeding of carbon monoxide absorption by physical activity and the possible influence of transient peaks of pollution. In the context of avoiding any hastening of exercise-induced angina, neither criticism of the existing standards has great substance. Although subjects of the key experiment sat in a car throughout exposure to the contaminated air, the dose of carbon monoxide was estimated from blood carboxyhaemoglobin rather than ambient carbon monoxide concentration readings; the threshold would thus have been similar if the subjects had been exercising. Likewise, substantial peaks of pollution were undoubtedly encountered on the freeway, yet results were similar to those obtained when exposures were repeated in the laboratory using uniform concentrations of carbon monoxide. The main concern with current standards is the degree of protection offered to the various high-risk groups discussed above. Further safeguarding of such individuals must be weighed against practical constraints; for example, Stewart et al. (1974) noted that in their studies of blood donors, 25 to 75% of supposed nonsmokers living in various United States cities already exceeded a 1.5% carboxyhaemoglobin.

INDUSTRIAL STANDARDS

We have noted already that earlier generations of industrialists accepted quite casually the exposure of employees to very high pressures of carbon monoxide; indeed, miners were expected to continue with their work until a mouse or a canary became virtually unconscious. When investigating the physiological responses to such hazards, it was necessary for J.S. Haldane to generate a carboxyhaemoglobin of 56% in his own blood (Douglas, 1963). As recently as 1948, one standard textbook of public health (Frazer and Stallybrass, 1948) was content to observe that "0.1 per cent (carbon monoxide) causes unpleasant symptoms, whilst exposure to 0.4 per cent causes death . . . . " We will here review possible limiting criteria and recommend a more acceptable industrial standard.

Possible Criteria

Some toxicologists have expressed exposure to hazardous gases and vapours in terms of a concentration-time product (CT, measured for example in $mg{\cdot}min{\cdot}m^{-3}$). Such an approach seems suited to a substance that can accumulate within the body, particularly if there is a likelihood of brief exposures to high concentrations of the agent; however, the CT index is misleading if the body becomes progressively saturated with the toxin, as is the case for CO. Some occupational exposures to carbon monoxide are of short duration, but if a simple concentration-time product were calculated for an entire eight-hour shift it would yield a misleadingly large index of exposure. There are two possible alternatives — to specify an ambient concentration that should not be exceeded and to specify a blood carboxyhaemoglobin level that should not be exceeded. The principles on which such specifications are based are similar to those for the urban environment, but the proposed safety margin is usually small or even nonexistent in an industrial situation (Hatch, 1972, 1973).

It is necessary to discuss and criticize current concepts, including the threshold limit value (TLV), the maximum acceptable concentration (MAC), and the emergency exposure limit (EEL).

Threshold Limit Value

For many years, a partial pressure of 10 Pa carbon monoxide was regarded as an acceptable TLV for United States industry (Katz and Frevert, 1928a; Gafafer, 1964). The first codification of acceptable standards can be traced to a pamphlet entitled "Occupational Hazards and Diagnostic Signs," published by the Metropolitan Life Insurance Company in 1918. Revision of the pamphlet was subsequently undertaken by the United States Department of Labor and more recently by the Division of Occupational Health, Public Health Service. The recommended threshold limit of 10 Pa carbon monoxide for an eight-hour exposure was reiterated in the 1964 edition of this manual (Gafafer, 1964). The 5 Pa limit can be traced to a meeting of the American Conference of Government Industrial Hygienists (ACGIH, 1971; Grut et al., 1970), which led to a "Documentation of Threshold Limit Values." The proposed standard exceeds even the maximum tolerable figure for urban exposure; indeed, the intent is to do no more than limit the blood carboxyhaemoglobin level to 10%.

The ACGIH (1971) argued that the 5 Pa standard would offer adequate protection to a healthy worker, although it admitted the proposed limits might not be sufficiently stringent to avoid what were termed "sensitive" responses, such as a deterioration of visual threshold or auditory duration discrimination. It was reasoned that such changes were not adverse with respect to health and that the corresponding carboxyhaemoglobin levels fell within the range normally encountered in smokers. More recently, the National Institute for Occupational Safety and Health (NIOSH, 1972) has recommended a further reduction of the TLV to 3.5 Pa. Standards of a similar order are accepted by many other countries (*see* Table XXVI).

The basic objective of the TLV is to define "conditions under which it is believed that nearly all workers may be repeatedly exposed day after day without adverse effects" (Schilling, 1973; D.M. Davies, 1975). Judgements concerning the safety margins for different industrial hazards have been influenced by the frequency and severity of ill effects, the potential for associated extramural exposure (in the present context, exposure to carbon

Table XXVI
OCCUPATIONAL EXPOSURE LIMITS (8-HOUR PERIOD)

Carbon monoxide pressure	Nations
5 Pa	Australia, Belgium, Canada, Finland, Switzerland, U.S.A., Yugoslavia
3.5	Sweden, Ontario
3.5 (max 20)	U.S.A. (1972-NIOSH proposed long-term goal)
3.0 (max 13)	East Germany
2.6	Hungary and Poland
2.6 (max 13)	Czechoslovakia
2.6 (max 4.4)	Rumania
1.7	Bulgaria, Russia

monoxide in the community), and the feasibility of proposed control measures. Determination of TLV is often based upon the "relationship between the prevalence of a disease and the level of exposure to a toxic substance" (Walford, 1973) rather than a consideration of the threshold for physiological responses. With many pollutants, the permitted industrial exposures far exceed allowable urban contamination. However, the premise has then been to move from a standard proven as safe in industry to an urban standard that was attainable and economically feasible. In the case of carbon monoxide, the urban standard in itself allows no more than a twofold safety margin even in healthy individuals at sea level, and there is no a priori reason for allowing workers to encounter a higher threshold dose than that on which the urban standard is based.

Critique of Threshold Limit Value

Several of the basic premises underlying the TLV are highly debatable. Foremost is the decision to protect only healthy workers. This may be a reasonable approach with some rarely encountered toxic materials, since sensitive individuals can then be screened from the labour force. However, in the case of carbon monoxide, a high proportion of employees in a wide variety of occupations are at risk, and many of the companies concerned lack a well-equipped medical service. Further, there are no simple test procedures to detect some of the numerically important bases for carbon monoxide sensitivity, such as clinically silent ischaemic heart disease (G.H. Epstein, 1973; B. Lown, 1979). Pregnant women form an important medical category with special vulnerability to carbon monoxide. Is it appropriate to regard such women as unhealthy, and if so, should provision be made for extended maternity leave in occupations such as tavern waitress, where a substantial exposure to carbon monoxide is likely? Unfortunately, TLVs are likely to be exceeded not in large companies that offer antenatal care and screen employees for susceptibility to carbon monoxide, but rather in small companies where much of the employment is on a part-time or casual basis. There are thus some grounds for suggesting that in the case of carbon monoxide the industrial threshold should be appropriate to protect high-risk populations. One counter argument is the absence of any safe threshold dose of carbon monoxide permissible for patients with cardio-respiratory disease. The zero threshold concept has some statistical justification (since extrapolation of dose/response curves to zero effect yields an exposure intercept that does not differ significantly from zero), but in practice a threshold dose of CO that protects the healthy adult against psychomotor disturbances and impairment of exercise tolerance will also offer adequate protection to most cardiac patients who are able to work. The decision concerning industrial versus urban standards is essentially an ethical problem (B.W. Dick, 1973; L. Goldberg, 1972). Does the individual who is enjoying the fruits of a worker's labour merit a larger margin of health protection than the worker himself or herself? Factors influencing such a decision include not only political realities such as the economic costs of reductions in urban and industrial carbon monoxide exposure but also the number of individuals at risk in a particular manufacturing process and judgements concerning the severity of body responses to various doses of carbon monoxide. Unfortunately, exposure to carbon monoxide occurs in a wide variety of occupations (*see* Table V), and there are many categories of employees with increased susceptibility to this agent (*see* Table XVI). Furthermore, the hazards of excessive exposure are severe, including an increase in the number of fatal heart attacks and a risk of fatal accidents. It is thus difficult to argue for any leniency of industrial controls. Lifting of the twofold safety margin relative to urban standards (above) may be an economic necessity, but suprathreshold exposure of the worker is certainly not warranted.

The observation that current carboxyhaemoglobin readings of employees are "no higher than in smokers" is a particularly specious argument against the reform of existing industrial standards. Cigarette smoking is an adverse life-style of the employee's choice, universally condemned by health experts; moreover, there is some evidence (*see* Chapter 5) that the carbon monoxide content of cigarette smoke (or the resultant blood carboxyhaemoglobin level) is sufficient to contribute to the development of atherosclerosis. Occupational health standards can hardly be based upon such a fallible criterion. It has also been suggested that sensitive responses be excluded from the industrial threshold on the basis that the physiological disturbances commonly discussed have no adverse effects upon health. However, the allowance of certain physiological disturbances would hardly help the industrialist, since there seems a close similarity between physiological and pathological thresholds in the case of carbon monoxide. Moreover, Hatch (1973) has suggested that all industrial air standards should be set below the limit of physiological compensation, thus leaving a safety margin for (1) unintentionally greater exposures, (2) impairment of the individual by age, intercurrent illness, or other types of environmental stress, and (3) possible cumulative effects from repeated exposure to the toxic agent. Sudden emergencies often lead to unexpectedly severe carbon monoxide poisoning, with little warning of toxicity prior to collapse of the worker. Vulnerability to CO may be increased by concurrent exposure to high altitude or an asphyxiating gas. Evidence on the cumulative effects of CO is somewhat equivocal, but certain data on the generation of atherosclerotic lesions (*see* Chapter 5) suggest such a pattern of response. In any event, the psychomotor disturbances reported at the lowest doses of CO can hardly be dismissed as trivial. From all that is known about cerebral hypoxia, a dose of carbon monoxide that is sufficient to cause impaired psychomotor function seems likely to increase the chance of accidents involving the worker, his colleagues, and expensive machinery. Even if the employee undertakes no dangerous mechanical processes while within the plant, there will be an increased risk of injury from an automobile accident during the homeward journey. Likewise, there are medical risks attached to an acute decrease of exercise tolerance, particularly in heavy industry. The cardiac work rate of the employee concerned is increased for the performance of any given task, and the likelihood of fatigue over the working day is augmented.

It is sometimes claimed that a smaller margin of safety is acceptable in industry than in the urban environment because the pressures of carbon monoxide in the work place are relatively constant and clearly defined. The average city dweller is certainly exposed to occasional pockets of carbon monoxide containing 10 to 100 times the monitored ambient concentrations. However, a similar increase of hazard is all too often encountered during industrial processes. Even in large and well-regulated plants, dangerous concentrations of gas accumulate not during normal operations but rather through mischance—the malfunction of a valve or the need to repair machinery. Dangerous exposures are even more frequent in small workshops, where both process and ventilation are highly variable. A substantial fraction of occupational episodes also involve vehicle exhaust; the dose of CO received by the subjects then depends upon the vagaries of both traffic density and weather conditions.

It has been suggested that because peak industrial concentrations of carbon monoxide are quite localized, the average employee encounters the worst measurable environment for only a short portion of the working day. A miner may walk through a pocket of carbon monoxide on his way to and from the coal face, while a blast furnace operator will encounter high carbon monoxide concentrations at certain key steps in the operating cycle of a furnace. Peak readings may thus overestimate the hazard. Again, the argument has some validity, but only in a limited number of occupations. In many situations, such as a parking tower or a customs booth, the attendant has little possibility of escape from the contaminated environment.

A further consideration is that industrial CO exposure is limited to a working day of eight hours or so. Given the much lower carbon monoxide pressures usual in the domestic environment, an employee may not reach full equilibrium with measured concentrations at the work site, and in any event there will be a good opportunity for reducing the body burden of carbon monoxide on returning home. In contrast, urban carbon monoxide pollution commonly has a periodicity of four to five days, allowing a complete equilibration with peak carbon monoxide pressures, with only a slow subsequent elimination of the gas. This historic distinction between urban and occupational exposure is now becoming blurred, and in a number of situations (for instance, spacecraft, submarines, underwater habitats, and arctic camps) twenty-four hour exposure to the working atmosphere is becoming the rule (D.M. Davies, 1975).

A final claim to specificity of the industrial standard is the influence of respiratory minute volume upon the uptake and clearance of carbon monoxide (Song and Eagan, 1978). A typical person spends most of the day in a sedentary state, the average energy consumption being no more than 15% above the basal figure. However, in a number of the industries where carbon monoxide is encountered (mining, firefighting, cargo handling, and the inspection of ships' holds, for example), relatively heavy physical work is required (Griggs, 1977). Given eight hours' exposure to 5 Pa of carbon monoxide, heavy work increases the absorbed dose of CO by 30%. With fifteen minutes of exposure to a peak of pollution, heavy work leads to a threefold increase in the absorbed dose. On such evidence, the permitted levels for heavy industry should be lower than for the general population, particularly with respect to transient peaks of pollution.

Excursion Values

The TLV is usually stated as a time-weighted eight-hour average. Excursions above the threshold are then permissible if there is a compensatory drop below threshold at some other point in the working shift. Nevertheless, it is specified that for substances in the concentration range 10 to 100 ppm (that is, a carbon monoxide partial pressure of 1 to 10 Pa), excursions should not exceed the mean value by more than 50% (that is, a maximum excursion to 7.5 Pa from a TLV of 5 Pa). Roach (1966) has queried the usefulness of such an arbitrary ceiling and has suggested a statistical basis for keeping exposures below the TLV, working from mean values. Samples are collected over a tenth of the biological half-time for the pollutant (that is, about twelve to fifteen minutes for carbon monoxide). If the maximum permitted gas-phase reading is kept 1 standard deviation (S.D.) below the published TLV, then the body burden for an eight-hour day is held 5 S.D. below this ceiling, allowing a substantial margin of safety. Although data sometimes depart from both a normal and a log-normal distribution (Shultz and Beauchamp, 1972), if twelve readings or fewer are available, the standard deviation can be approximated by the formula (S.A. Hall, 1973)

$$\text{S.D.} = \text{range } (n-1)^{1/2}$$

where n is the available number of samples.

Brief Exposures

The WHO (1979) has collated other proposed standards for short exposures (*see* Table XXIV), including a ten-minute excursion value of 10 Pa allowed by the Pennsylvania Department of Health and an upper limit of 20 Pa proposed by NIOSH (1972). Such figures are alarmingly high relative to current knowledge of carbon monoxide poisoning; indeed the first would cause a 14% carboxyhaemoglobin reading.

Some nations have specified *maximum allowable concentrations* (MAC) that should in no circumstances be exceeded. The Russian standard of 1.7 Pa is in essence an MAC and as noted above is based upon the avoidance of physiological and psychological disturbances rather than the risk of developing adverse clinical effects (International Labour Office, 1970; Hartmann, 1972). *Emergency exposure limits* (EEL) also find application in estimating the risks from large scale accidental air pollution and in assessing the hazards encountered by rescue workers; EELs are commonly proposed for ten-, thirty- and sixty-minute exposures (Ziethuis, 1970).

Use of Blood Samples

Local peaks of gas concentration, variable shift lengths, and substantial differences of respiratory minute volume make it difficult to regulate industrial contamination in terms of ambient carbon monoxide pressures (*see* Chapter 2). For example, a fifteen-minute exposure of a stevedore to a gas pressure of 20 Pa could yield a blood carboxyhaemoglobin of 5.2%, although the eight-hour time-weighted average exposure might still not exceed the apparently harmless figure of 0.7 Pa. If TLVs and EELs are to be used effectively in the work place, they must therefore be industry or even task specific.

For such purposes as enforcing air quality legislation and preventing excessive carbon monoxide exposure, there is much to commend either direct blood sampling of affected workers or the making of bloodless estimates of carboxyhaemoglobin levels (*see* Chapter 2). Ambient gas monitors can then be used as a second line of defence, serving to regulate the ventilation of tunnels and mines and to provide immediate warning of emergencies.

Nevertheless, maximum permitted carboxyhaemoglobin values should be reduced from the currently accepted figures of 10% for an eight-hour day (ACGIH, 1971) or 14% for a ten-minute excursion (Pennsylvania Department of Health, 1972); indeed, it seems undesirable that the blood carboxyhaemoglobin of the nonsmoking worker be allowed to exceed 3%.

High-Risk Workers

Stewart (1974) and Jones and Knelson (1980) have argued that since any dose of carbon monoxide has some adverse effect upon the patient with chronic cardiac disease, it is impossible to set a TLV to protect this segment of the labour force. The alternatives for a regulatory agency are (1) to accept this counsel of despair, (2) to set what may prove a costly and excessively cautious standard, or (3) to require a medical examination that would exclude employees with high-risk conditions. The last option implies provision (possibly through an increase of specialist training in occupational health) for the screening of a substantial segment of the working population.

Pregnancy is usually well recognised in the second and third trimesters, and the adverse effects of carbon monoxide upon foetal oxygen supply (R. Davie et al., 1972; Goldstein, 1971; Longo, 1977; Fechter and Annau, 1977; Prigge and Hochrainer, 1979; E.N. Wood, 1979; Schwetz et al., 1979) seem likely to be most severe in the final three months of gestation. On the other hand, there is a greater risk of provoking abnormalities of development if foetal nutrition is impaired during the first trimester of pregnancy (Longo, 1977); this implies the desirability of keeping all pregnant employees away from sources of carbon monoxide.

Screening for the other conditions that may increase susceptibility to carbon monoxide (*see* Table XVI) follows the normal principles of preventive medicine. For example, if automated measurements of vital capacity and forced expiratory volume disclose significant restrictive or obstructive pulmonary disease (Aronow et al., 1977), the resultant limitation of oxygen transport should be explored further in the clinical laboratory, including measurement of the arterial oxygen pressure at the level of activity anticipated during normal daily work and a treadmill assessment of exercise tolerance. Likewise, if a simple blood haemoglobin reveals significant anaemia, laboratory tests should be carried out to exclude haemolytic disease.

The increased risks of CO poisoning associated with urban driving and cigarette smoking have been discussed above. In a number of occupations, the adverse effects of carbon monoxide exposure are compounded by a low partial pressure of inspired oxygen, for example, when driving a steam locomotive through mountain tunnels (Haldane and Priestley, 1935), flying a light aircraft (McFarland et al., 1944), fighting fires, or patrolling mountain tunnels on busy highways. In all of these situations an altitude-specific ceiling of carbon monoxide exposure becomes appropriate (K. Master, 1976; Mitchell et al., 1979; Weiser et al., 1980).

In some industries, the toxicity of carbon monoxide may be enhanced by simultaneous exposure to other pollutants such as phenol and dust (Elfimova et al., 1972), NO_2 (Busey, 1972; Nakajima et al., 1972), ozone (S.D. Murphy, 1964), and benzene (Ponsold et al., 1978). Likewise, susceptibility to CO may be increased by ingestion of alcohol (Stewart et al., 1974; Rockwell and Weir, 1975; Pankow et al., 1974) or depressant drugs (Montgomery and Rubin, 1971).

CONTROL MEASURES

Urban Architecture

Since the main source of urban carbon monoxide is the automobile, the key to control is to reduce the number of petrol-driven vehicles entering large cities (*see* Table XXVII). To some extent this can be accomplished by legislation (for example, by requiring costly permits to enter the city core or by restricting parking facilities to drivers who are carrying at least two passengers). However, a more effective approach is to design the car out of the city.

Table XXVII
EMISSIONS OF CARBON MONOXIDE IN A CITY OF ABOUT 2 MILLION (TORONTO, 1973-74)*

Source	Emissions of CO (tons $\times 10^3 \cdot yr^{-1}$)
Automobiles	270.3
Other transportation	1.5
Space heating	1.2
Thermal power	2.1
Other industries and utilities	104.3
Total	379.4

*Based on data of Economic Council, 1974.

If a new town is being planned, each quadrant may be arranged as a self-contained unit, with houses, schools, shops, offices, and appropriate light industry linked by pleasant footpaths and cycleways but vehicles restricted to the outer edge of the quadrant. Thus Martin and Oeter (1978) commented "the most (energy) efficient pattern is one of small to medium-sized rectangular or concentric settlements, arranged in polynuclear fashion . . . urban sprawl must be less energy efficient than any planned approach . . . medium-sized towns or urban districts (25,000 to 100,000 population) offer energy conservation advantages over smaller and larger centres." In the larger cities, an efficient mass transit system is a necessity. The cleanest solution is a network of subway routes (Shephard, 1974); these have a high capacity (about 50,000 passengers per hour) but need costly subsidies unless there is high-density office and apartment construction overhead. Buses, trolleybuses, and street cars offer a cheaper and lower-capacity alternative; their acceptability has been substantially increased by the introduction of reserved bus lanes and priority operation of traffic signals by buses and street cars.

In general, narrow streets favour the accumulation of carbon monoxide, and the best plan for the historic core of older cities is the creation of a pedestrian precinct or mall (Wright et al., 1974). When this was tried in Toronto, there was little evidence of increased pollution on surrounding streets. Nevertheless, if the arrangement is permanent, access roads and parking towers can often be created to replace slums ringing the core. Vehicle speed is an important variable, and gains can thus result from such traffic engineering measures as turning lanes, advanced-turn signals, no left turn signs, and one-way street systems (Wright et al., 1974). A combination of enormous road widths and high vehicle speeds leads to very low carbon monoxide levels on superhighways. If such routes lead directly to a subway parking area, they provide a useful means of moving the more outlying commuters. However, if the highway continues to the centre of a town, it merely serves to increase the number of polluting vehicles entering the most congested areas.

A few other suggestions may be made with regard to building and highway design. Tall buildings should be set back a reasonable distance from the curb. An extra metre of sidewalk does much to reduce the exposure of the pedestrian. Nevertheless, where possible the walker should be further protected from traffic by trees and other barriers. An ideal arrangement is to place the walkway behind the building. Overhanging entrances should be avoided, and where such structures already exist waiting vehicles should be asked to turn off their engines. Long underpasses should be avoided or provided with artificial ventilation. Where possible, shorter underpasses and cuttings should be arranged to take advantage of the ventilation offered by the prevailing wind.

Vehicle Design

The potential for reducing emissions of carbon monoxide by a change of vehicle design has been discussed in Chapter 1. Unfortunately, many of the possible remedies have other adverse environmental effects.

Perhaps the simplest remedy, already popular because of the energy crisis, is to drive smaller and lighter cars. Since less gasoline is then consumed per vehicle-kilometer, the output of carbon monoxide inevitably falls. An argument against this solution sometimes put forward by the North American manufacturers of large cars is that the chance of survival in a highway crash is reduced in a small vehicle. However, since any collision is usually with another car, the risks associated with disproportionate inertia should disappear as current North American automotive dinosaurs die of thirst. Moreover, if automobile manufacturers were sincere in their concern about collisions, they could add to small cars such devices as inflatable bumpers that would readily match the protection currently afforded by the cab of a larger car.

The second simple change would be to insist that a car's exhaust pipe be mounted on the roof (as in some trucks and buses); the carbon monoxide, which is lighter than air, would then escape upwards, avoiding both pedestrians and the occupants of following vehicles.

Existing emission controls have done much to curb the production of carbon monoxide, although as we have already noted the catalytic afterburner may increase the atmospheric burden of sulphuric acid aerosol (Maugh, 1977). A proportion of cars are also switching to diesel fuel, which greatly reduces carbon monoxide production at the expense of other unpleasant smoke constituents. The ultimate solution will probably be electrically powered cars, and if the electricity is generated from renewable sources, the change should have no adverse environmental effect.

Industrial Processes

In most industrial situations, there is a trade-off between the risks of carbon monoxide exposure and the costs of preventive or control measures (Goldberg, 1972; Duck, 1973). Options include (Haag, 1979) (1) a change of process, (2) control of the pollution at source, (3) an increase of general ventilation, and (4) the use of personal protective equipment, with education of the vulnerable worker in all aspects of prevention and treatment of poisoning (N.J. Smith, 1975).

Change of Process

Many examples of change of process could be sited. Battery operation has been adopted for the towing of aircraft and for forklift trucks that are operated in confined spaces. Diesel engines have replaced gasoline engines for many commercial purposes. Electric traction has been adopted on both interurban and underground railway systems. Multiple-chamber refuse burning has replaced the open burning of refuse, decreasing carbon monoxide output from 50 g per kg of waste to near zero. Natural gas has replaced producer and water gas as an industrial and heating fuel, or alternatively odorants have been added to warn of leakage (Rubino et al., 1970). Electrically powered conveyor belts now move cars through inspection lanes, and similar devices have been proposed for congested tunnels.

In many instances, carbon monoxide accumulates through an incomplete combustion of carbon fuels, and such inefficiency is particularly undesirable at a time of diminishing energy reserves. Smith and Gruber (1966) commented that whereas electrical generators emitted 0.5 tons of carbon monoxide per ton of coal that was burned, in some commercial plants the figure was as high as 50 pounds per ton.

Control at Source

Control at source involves an appropriate manipulation of the four main variables influencing carbon monoxide production (fuel/air ratio, flame temperature, gas residence time, and turbulence in the combustion chamber). Thus an increased air supply reduces the output of carbon monoxide from a foundry cupola (Draper et al., 1968). Some installations produce more than 1000 kg of carbon monoxide per ton of pig iron, but improved methods of producing iron and steel can cut emissions to less than 10 kg per ton. In petrol-driven vehicles, emissions can be moderated by use of a leaner fuel, with better mixing of air and gasoline in the carburetor, preheating of intake air, an increase of idling speed, a retardation of ignition, improvements in cylinder head design, the incorporation of thermal reactors into the exhaust chamber, the use of oxidizing or reducing catalysts, the introduction of secondary air systems, and the recycling of exhaust (National Academy of Sciences, 1973). Other practical measures include turning off engines in vehicles waiting at airports and bus terminals (Howlett and Shephard, 1973) and improvements of highway design. Unfortunately, automotive designers have little intrinsic interest in the carbon monoxide that is produced by their vehicles since complete combustion of CO would improve fuel efficiency by no more than 0.5%, a gain that is more than dissipated in devices such as afterburners. Nevertheless, the potential for design improvement is dramatically illustrated by the legally required decreases in the emissions from United States light-duty automobiles (Stern et al., 1973; Walsh and Nussbaum, 1978) from 39 g per vehicle-mile in the late 1960s to 15 g per mile in 1976, 7 g per mile in 1980, and 3.4 per mile after 1981.

Ventilation

If a particular area of the work place is known to be heavily contaminated, appropriate ventilation must be arranged (Roach, 1973). A *steady air stream* may be used to carry contaminated air away from a known source of carbon monoxide or to bring a supply of fresh air through a ventilation grille. In either case, the stream becomes larger and more slowly moving as it mixes with surrounding air. Contamination of a room is greatly reduced if a hood is placed over the source of carbon monoxide, particularly if the exhaust ducting is fitted with a fan of adequate power.

Random turbulent mixing occurs if the carbon monoxide is emitted continuously and at low velocity from a point source. *Dominant air movements* can be observed near windows, doors, and exhaust hoods. The velocity of air movement towards such openings is inversely proportional to the distance from the openings. A steady state is reached when the emission of carbon monoxide is balanced by removal through the various available forms of ventilation. Given K air changes per hour, the half-time is 0.693/K hours. With a traditional figure for K (six air changes per hour), the half-time is only 6.7 minutes; however, it can become much longer if a room is made more airtight in order to conserve fuel. After a steady state has been reached, pollutant levels are proportional to the inverse square of the distance from the source.

Natural ventilation does not offer satisfactory protection against a highly toxic gas such as carbon monoxide. The extent of such ventilation inevitably varies greatly with wind pressure, wind direction, the extent of solar radiation, indoor convection currents, and the personal whim of the workers (opening or closing of windows and doors). The alternative is *dilution ventilation,* where air is forced into or from a room by a fan. The major disadvantage of this approach is the cost of heating incoming air in the winter and cooling it in the summer.

In the absence of carbon monoxide, the minimum requirement is 0.4m^3 of fresh air per person per minute or (in a 20m^3 room) one air change per person every fifty minutes; when heavy physical work is undertaken, a doubling of this standard is recommended. If carbon monoxide is present, the required ventilation can be calculated from the formula for pollution of a hemisphere.

$$C_A = 0.1\ EB^2\ /\ VA^2$$

where C_A is the partial pressure of carbon monoxide (Pa) at a distance A metres from the source, E is the emission of carbon monoxide ($cm^3 \cdot sec^{-1}$), V is the rate of exchange of pollutant between the hemisphere and the uncontaminated surroundings ($m^3 \cdot sec^{-1}$), and B is the radius of the hemisphere in metres (commonly set at the mean distance from the pollutant source to the centres of the walls and ceiling of the room). If the source of carbon monoxide is close to one wall, pollutant levels are twice as large (in effect, pollution of a quarter hemisphere), while if it is in a corner, levels are four times as large.

If doors and windows are closed, the natural ventilation of a small room is often in the range 5×10^{-4} to 5×10^{-5} $m \cdot sec^{-1}$ ($m^3 \cdot sec^{-1}$ per m^2). Setting the allowable value of C_A at 5 Pa and putting A = 1 metre and $V = 5 \times 10^{-5} m \cdot sec^{-1}$, the maximum emission of carbon monoxide is only 0.9 $ml \cdot min^{-1}$ if the source is at the centre of the room and 0.2 $ml \cdot min^{-1}$ if it is located in a corner. Dilution ventilation can cope with emissions at least 100 times as large, although it is important to ensure that the intake air is reasonably free of carbon monoxide. If there are several sources of carbon monoxide, as when cigarette smokers are scattered about a room, the likely distribution of carbon monoxide can be calculated graphically by adding estimates for individual hemispheres. With more than five sources, the air contamination becomes relatively uniform, and the formula simplifies to

$$C_A = 0.1\ E/V\ (Pa)$$

Local ventilation is the appropriate method of controlling the very high concentrations of carbon monoxide encountered in some industrial situations. Possible techniques include a hood, a slot, and a booth. A hood must induce an air movement greater than the dispersion velocity from the source, so that all of the contaminant is extracted. Given a flow rate of V $m^3 \cdot sec^{-1}$, the local air velocity V_A at a distance A from the hood is $V / 4\Pi A^2 m \cdot sec^{-1}$. Provided that the carbon monoxide is emitted at low velocity, a hood will serve its purpose until the extraction velocity drops below the velocity of local air currents; setting the latter at 0.1 $m \cdot sec^{-1}$, we find A = 0.9 V metres. Similar calculations limit the effective distance to 1.3 V metres for a wall opening and 1.6 $V \cdot L^{-1}$ metres for a narrow slot of length L metres. In practice, somewhat higher exhaust velocities (or closer placements of the hood) are necessary to counter additional air currents generated by the occupants of the room, machines, draughts, and the industrial process itself. If vertical access to the work area is necessary, the hood is replaced by an exhaust slot mounted around the periphery of the machine; if the area of the emitting surface is less than 1.5m^2, the slot provides more efficient control of contaminants than a hood. A third possible option is to enclose the machine or process in a booth that is open on one side and is exhausted from the other; an exhaust flow of 0.5 $m \cdot sec^{-1}$ usually suffices with this type of arrangement.

The *jet stream* from the industrial process may contribute to ventilation. Given a source diameter of C metres, the linear flow rate V_A at a distance of A metres from the source is $3C(V_E)/A$ $m \cdot sec^{-1}$, where V_E is the jet velocity. The diameter of the cone of polluted air is A/2, and carbon monoxide will not escape into the room provided that V_A exceeds local air currents (0.1 $m \cdot sec^{-1}$), the hood diameter is larger than A/2, and the rate of exhaustion of the hood exceeds the volume flow of contaminated air ($0.3A(V_o)/C$, where V_o is the volume flow rate at the jet).

The *energy cost* associated with forced ventilation is substantial (Martin and Oeter, 1978). It can be calculated from the temperature differential $\Delta°C$ between indoor and outside air, using the formula

$$\Delta° C = 0.0144\ H/V$$

where H is the heat output in $kJ \cdot min^{-1}$, and V is the ventilation in $m^3 \cdot sec^{-1}$. On a severe winter day, $\Delta°C$ may rise to 40° C, so that the additional ventilation needed to control cigarette smoke accumulation in a room occupied by six smokers ($5.4 m^3 \cdot min^{-1}$; Shephard 1982a) costs 250 $kJ \cdot min^{-1}$, or 4.16 kWatt. A typical industrial ventilation ($0.5 m \cdot sec^{-1}$ through a 1 $metre^2$ hood) would give a heat loss of 23.1 kWatt in similar circumstances. The alternative tactic would be to scrub and recirculate room air, but unfortunately there is no simple way of removing large quantities of carbon monoxide from the air stream. The most promising methods of reducing energy losses are to add a heat exchanger to the ventilation system and to emphasize local rather than general ventilation as a means of eliminating carbon monoxide.

Protective Equipment

In some occupations such as mining and firefighting, potential exposures to carbon monoxide are so severe that personal protective equipment must be provided (American Industrial Hygiene Association, 1963; British Standards Institute, 1969). Long air hoses are helpful in some instances but severely limit the mobility of personnel. The quantity of carbon monoxide emitted is usually rather large for the gas to be removed by a personal filter, although an operational life of sixty minutes in 1.5% carbon monoxide is claimed for a German design of respirator where dry air is passed through a bed of Hopcalite, which catalyzes the conversion of carbon monoxide to carbon dioxide (Place, 1970). Rescue workers are commonly issued with a closed-circuit or pendulum type of breathing equipment (Shephard, 1982b), oxygen being added from a small and portable high-pressure cylinder in order to maintain the volume of the system. Closed circuits have the advantage that they are effective in atmospheres containing high percentages of carbon dioxide and low percentages of oxygen (a common occurrence in fires). Other systems are based on the carrying of a compressed air cylinder (Loke et al., 1976; Radford and Levine, 1976).

A surprisingly large amount of training is necessary to ensure effective use of respiratory equipment. The main danger with both self-contained and line-operated devices is an inward leakage of contaminated air. This can occur if the mask is not fitted carefully to the face or an expiratory valve becomes jammed in a partially opened position. If the ambient carbon monoxide pressure is 5 kPa, a leakage of one part in a thousand causes the circuit contamination to exceed the TLV. Moreover, in a closed or partial rebreathing system, leakage has a cumulative effect. It is particularly difficult to ensure that a facemask does not become dislodged during arduous physical operations such as mine rescue. Failure of the breathing system may also develop if the peak air or oxygen flow to the mask fails to match the ventilatory demands of heavy work. In such circumstances, the worker has little option but to aspirate contaminated air around the rim of the mask. Finally, with line-operated systems, a worker may encounter a dangerous pocket of carbon monoxide while walking between "plug-in" points.

Haldane (1915) commented that workers should be trained in the use of breathing equipment "under conditions which as regards muscular work, heat, smoke, poisonous air and other adverse conditions would be worse than ever likely to be met underground."

Attempts to introduce self-contained breathing units into blast furnace and coke oven operations have not been too successful (J.G. Jones, 1961). Workers complained that the devices provided were too heavy, too bulky, and too complicated. The wearing of a noseclip and mouthpiece was accepted by only a few employees, while a full facepiece was said to be excessively hot, causing claustrophobia, and impeding vision (especially ability to perform fine work). British steelworks currently supply workers with a light oronasal mask supplied from a compressor or a cylinder of compressed air. A good flow of air not only avoids inward leakage of carbon monoxide but also serves to cool the face.

Treatment Facilities

If there is a risk of industrial carbon monoxide poisoning, it is important that employees are trained to recognise symptoms and to institute early and appropriate first aid treatment.

If a worker must descend into a contaminated tank or other enclosed space, a belt and lifeline should be worn so that he or she can be hauled out in the event of failure of breathing equipment. The first concern in treatment is to move the victim to a respirable atmosphere, but in a mine where much contamination has occurred it may be necessary to provide oxygen (with or without a small concentration of carbon dioxide) while being brought to the surface on a stretcher. In view of the probable loading of the oxygen transport system, further taxing of the heart by exertion should be avoided. Even if recovery appears complete, it is important that the worker be driven home and observed carefully for several days. All rescue workers and as many as possible of the general labour force should also be taught the basic principles of cardiopulmonary resuscitation, so that early care can be given to those with more severe poisoning.

Blood sampling is desirable on both medical and legal grounds. A careful history of the accident may be helpful in modifying safety rules to prevent a recurrence. We have noted (*see* Chapter 3) the value of hyperbaric oxygen in speeding the elimination of carbon monoxide, and if poisoning is severe, the medical officer in charge of the case should consider evacuation of the patient to the nearest hyperbaric facility.

LEGAL IMPLICATIONS

There have been occasional incidents where forensic experts have needed to distinguish suicide from murder involving the use of carbon monoxide (Kerr, 1946). However, the physician is more commonly involved in litigation involving industrial exposure and vehicle accidents. There have also been a substantial number of legal battles over the accumulation of cigarette smoke in public places.

Industrial Exposure and Vehicle Accidents

The principles of investigation are similar in the two types of situation. In some industrial emergencies, the exposure to carbon monoxide is sufficiently overwhelming that there is little reason to dispute the cause of death. However, if an aircraft pilot or a car driver is found dead following a crash, it becomes arguable that the accident was not related to carbon monoxide poisoning but rather that carbon monoxide accumulated subsequent to the crash due to such factors as fracture of an exhaust pipe or a structural fire.

Key factors in determining etiology include the following:

1. A careful personal history, taken from family and verified by independent witnesses. This should cover smoking, drinking, and drug habits, driving history (duration, mileage, and traffic violations), and information on psychiatric state. The crash could represent a suicide, and likewise carbon monoxide could have accumulated during an attempted suicide (running the car in a closed garage);
2. A determination of blood carboxyhaemoglobin level, relating the observed figure to that needed for death, loss of consciousness, and disturbance of psychomotor function. If the sample was collected postmortem, the analyst

must exclude distortion of pigment readings by bacterial formation of methaemoglobin and nitrohaemoglobin;

3. A postmortem examination of the victim to determine the severity of wounds sustained, the likely period of survival, and thus the potential for inhaling carbon monoxide after the crash; and

4. A full mechanical examination of the vehicle and surrounding roadway. Was there any mechanical cause for the crash? Did the victim attempt to brake and if so at what distance from the obstacle? Were driving patterns normal prior to the emergency? Was any leakage of the exhaust system present before the crash? Were the windows open or closed? Was the air blower switched on? Had the driver or other vehicle occupants been smoking?

It is sometimes alleged that industrial exposure contributes to a fatal or a nonfatal heart attack. Possible development of atherosclerosis and other chronic symptoms due to frequent blood carboxyhaemoglobin readings of 10% and more is still a matter of vigorous debate. A more likely situation for a compensation claim is the discovery of a high carboxyhaemoglobin reading during intensive care; for example, the author was involved in one case where it was alleged the blood carboxyhaemoglobin was still 15% after thirty-six hours of treatment in an oxygen tent. Key questions in this discussion were an obvious source of exposure (the employee had been repairing a snowmobile during winter, the doors of the workshop were closed, and no auxiliary hosepipe had been fitted to the exhaust), symptoms beginning while at work (most courts are reluctant to admit a cause-and-effect relationship in the absence of bridging symptoms), and the high carboxyhaemoglobin recorded in the hospital notes. In fact, the value is suspiciously high if oxygen had been administered continuously for thirty-six hours, as the hospital record indicated. The family admitted the victim had been a smoker; unfortunately, the hospital had failed to record smoking habits subsequent to admission, although the presumption was that smoking had been precluded by use of the oxygen tent.

Claims of lasting cerebral damage following nonfatal carbon monoxide poisoning must be viewed with some skepticism, particularly if blood carboxyhaemoglobin readings when first rescued are less than 40%. As noted earlier, Haldane was himself able to withstand repeated blood carboxyhaemoglobin readings of over 50%, apparently without prejudice to his intellect. The author, likewise, has experience of a case of shallow water blackout where the victim was unconscious in the pool for two-and-one-half minutes, but residual cerebral effects lasted only two to three hours; if consciousness is lost, some lasting cerebral disturbance is possible, but if the victim is conscious when rescued, lasting abnormalities seem unlikely.

Cigarette Smoking

Although the accumulation of cigarette smoke can occasionally cause quite a substantial build up of carbon monoxide (*see* Chapter 7), the main bases for legislation against smoking in public places are the fire hazard, annoyance by irritant components of the smoke, and a possible risk of carcinogenesis (Pollard and Brennan, 1978).

Until recently, many authors believed that the prospect of enacting effective control legislation was negligible (R.E.W. Fisher, 1962; Maxwell, 1962; James, 1964; Long, 1964; Neuberger, 1963). Many of the courts also apparently favoured the smoker. For instance, in 1971, a German taxi driver was fined 50 D-marks for telling a passenger rather bluntly not to smoke inside his vehicle (Eberle, 1977). However, the tide of public opinion has changed over the past decade. In 1974, a second dispute arose in a West German taxicab, and on this occasion the court ruled in favour of the nonsmoking vehicle operator (Eberle, 1977). A key factor in the revised decision was new evidence that the driver's health could indeed have been endangered by the passenger's cigarettes. The following year, a dentist brought a successful action against the Schleswig-Holstein Dental Association for unwilling exposure to cigarette smoke at a legally required continuing education course (Medizinische Monatschrift, 1975). In reaching judgement, the court ruled the appellant was entitled to freedom from harm or inconvenience unrelated to the prime purpose of the event and adverse health effects from the smoke exposure could not be excluded. An action brought concerning smoke contamination of the Louisiana Superdome stadium was dismissed on the basis that the patron had voluntarily decided to attend the stadium and thus had no constitutional right to privacy (that would include unpolluted air). In contrast, a case of smoke exposure at work was successful, since the plaintiff was obliged to remain within the polluted area during normal working hours and presented medical evidence of allergy to cigarette smoke (Pollard and Brennan, 1978; Shimp et al., 1976).

Legislation restricting smoking in public places has now been proposed in forty-five of the fifty United States, exceptions being jurisdictions where tobacco growing makes a major contribution to the economy (Vanderslice, 1976). Nevertheless, only about half of the states concerned have yet enacted these proposals (National Clearing House, 1976, 1977; National Center for Disease Control, 1978). Further details on voluntary and mandatory restriction of cigarette smoking are presented elsewhere (Shephard, 1982a).

REFERENCES

Abbatiello, E.R., and Mohrman, K.: Effects on the offspring of chronic low exposure to carbon monoxide during mice pregnancy. *Clin Toxicol, 14:* 401-406, 1979.

Abramson, E., and Heyman, T.: Dark adaptation and inhalation of carbon monoxide. *Acta Physiol Scand, 7:* 303-305, 1944.

American Conference of Governmental Industrial Hygienists: *Documentation of the Threshold Limit Values for Substances in Workroom Air,* 3rd ed., ACGIH, 1971, pp. 41-43.

Adair, G.S.: The hemoglobin system. VI. The oxygen dissociation curve of hemoglobin. *J Biol Chem, 63:* 529-545, 1925.

Adams, J.D., Erickson, H.H., and Stone, H.L.: Myocardial metabolism during exposure to carbon monoxide in the conscious dog. *J Appl Physiol, 34:* 238-242, 1973.

Agostini, J.C., and Ramirez, R.G.: Successful reversal of lethal carbon monoxide intoxication by total body asanguineous hypothermic perfusion. *Surgery, 75:* 213-219, 1974.

Air Pollution Control Association: *Air Quality Criteria for Carbon Monoxide.* Washington, D.C., U.S. Govt. Print. Office, (DHEW Publication AP-62), March 1970.

Allen, T.H., and Root, W.S.: Colorimetric determination of CO in blood by an improved palladium chloride method. *J Biol Chem, 216:* 319-323, 1955a.

Allen, T.H., and Root, W.S.: Colorimetric determination of CO in air by an improved palladium chloride method. *J Biol Chem, 216:* 309-317, 1955b.

Altland, P.D., and Rattner, B.A.: Effects of nicotine and carbon monoxide on tissue and systemic changes in rats. *Environ Res, 19:* 202-212, 1979.

Altshuller, A.P., Ortman, G.C., Saltzman, B.E., and Neligan, R.E.: Continuous monitoring of methane and other hydrocarbons in urban atmospheres. *J Air Pollut Control Assoc, 16:* 87, 1966.

Alvis, H.J., and Tanner, C.W.: Carbon monoxide toxicity in submarine operations. *Arch Ind Hyg Occup Med, 6:* 404-406, 1952.

Ambient air quality criteria. *Ontario Gazette, 107-48:* 4739-4741, 1974.

American Heart Association: *Exercise Testing and Training of Apparently Healthy Individuals: a Handbook for Physicians.* New York, American Heart Association, 1972.

American Industrial Hygiene Association: *Respiratory Protective Devices Manual.* American Industrial Hygiene Association, 1963.

Amerson, S.S.: Poison in cigarettes. *JAMA, 49:* 625, 1907.

Amiro, A.G.: Carbon monoxide presents public health problem. *J Environ Health, 32:* 1969-70.

Ammundsen, E., and Grut, A.: Determination of carbon monoxide in the blood. *Acta Med Scand, 115:* 151-162, 1943.

Anderhub, H.P., Hofer, P., and Scherrer, M.: Normal-warte der Hb-CO Sättigung des Blutes. *Schweiz Med Wschr, 100:* 739-745, 1970.

Andersen, K.L., Masironi, R., Rutenfranz, J., and Seliger, V.: *Habitual Physical Activity and Health.* Copenhagen, World Health Organization Regional Office for Europe, 1978.

Anderson, D.O.: The effects of air contamination on health. Parts II and III. *Can Med Assoc J, 97:* 585-593, 802-806, 1967.

Anderson, D.O.: Problems created for ice arenas by engine exhaust. *Am Ind Hyg Assoc J, 32:* 790-801, 1971.

Anderson, E.W., Strauch, J., Knelson, J., and Fortuin, N.: Effects of carbon monoxide (CO) on exercise electrocardiogram (ECG) and systolic time intervals (STI). *Circulation, 44, Suppl. II:* 135, 1971.

Anderson, E.W., Andelman, R.J., Strauch, J.M., Fortuin, N.J., and Knelson, J.H.: Effect of low-level carbon monoxide exposure on onset and duration of angina pectoris. A study in ten patients with ischemic heart disease. *Ann Intern Med, 79:* 46-50, 1973.

Anderson, G., and Dalhamn, T.: Hälsoriskerna vid passiv rökning. *Lakartidningen, 70:* 2833-2836, 1973.

Anderson, G.K.: Treatment of carbon monoxide poisoning with hyperbaric oxygen. *Milit Med, 143:* 538-541, 1978.

Annau, Z.: The comparative effects of hypoxic and carbon monoxide hypoxia on behavior. In Weiss, B., and Laties, V.G. (Eds.): *Behavioral Toxicology*. New York, Plenum Publishing Corp., 1974, pp. 105-107.

Anthony, A., Ackerman, E., and Strother, G.K.: Effects of altitude acclimatization on rat myoglobin. Changes in myoglobin content of skeletal and cardiac muscle. *Am J Physiol, 196:* 512-516, 1959.

Antonini, E.: Hemoglobin and its reactions with ligands. *Science, 158:* 1417-1425, 1967.

Antweiler, H.: Physiologische Kinetik der CO-inhalation. In Berlin, A., and Smeets, J. (Eds.): *Proceedings. Health Effects of Carbon Monoxide Environmental Pollution.* Luxembourg, Commission of European Communities Health Protection Directorate, 1974, pp. 197-207.

Apthorp, G.H., Bates, D.V., Marshall, R., and Mendel, D.: Effect of acute carbon monoxide poisoning on work capacity. *Br Med J, (ii):* 476-478, 1958.

Armitage, A.K., Davies, R.F., and Turner, D.M.: The effects of carbon monoxide on the development of atherosclerosis in the white carneau pigeon. *Atherosclerosis, 23:* 333-344, 1976.

Armstrong, H.E.: Carbonic oxide in tobacco smoke. *Br Med J, (i):* 992-993, 1922.

Aron, R.H., and Aron, I.M.: Statistical forecasting models. Part I. Carbon monoxide concentrations in the Los Angeles basin. *J Air Pollut Control Assoc, 28:* 681-684, 1978.

Aronow, W.S.: Carbon monoxide and cardiovascular disease. In Wynder, E.L., Hoffman, D., and Gori, G.B. (Eds.): *Modifying the Risk for the Smoker. Vol. I. Proceedings of the 3rd World Conference on Smoking and Health.* Washington, D.C., U.S. Govt. Print. Office (DHEW Publication NIH 76-1221), 1976, pp. 321-328.

Aronow, W.S.: Effect of passive smoking on angina pectoris. *N Engl J Med, 229:* 21-24, 1978.

Aronow, W.S., Dendinger, J., and Rokaw, S.N.: Heart rate and carbon monoxide level after smoking high-, low-, and non-nicotine cigarettes: a study in male patients with angina pectoris. *Ann Intern Med, 74:* 697-702, 1971.

Aronow, W.S., and Rokaw, S.N.: Carboxyhemoglobin caused by smoking non-nicotine cigarettes. Effects on angina pectoris. *Circulation, 44*: 782-788, 1971.

Aronow, W.S., Harris, C.N., Isbell, M.W., et al.: Effect of freeway travel on angina pectoris. *Ann Intern Med, 77*: 669-676, 1972.

Aronow, W.S., and Isbell, M.W.: Carbon monoxide: effect on exercise-induced angina pectoris. *Ann Intern Med, 79*: 392-395, 1973.

Aronow, W.S., Cassidy, J., Vangrow, J.S., et al.: Effect of cigarette smoking and breathing carbon monoxide on cardiovascular hemodynamics in anginal patients. *Circulation,50*: 340-347, 1974.

Aronow, W.S., Stemmer, E.A., and Isbell, M.W.: Effect of carbon monoxide exposure on intermittent claudication. *Circulation, 49*: 415-417, 1974.

Aronow, W.S., and Cassidy, J.: Effect of carbon monoxide on maximal treadmill exercise: a study in normal persons. *Ann Intern Med, 83*: 496-499, 1975.

Aronow, W.S., Ferlinz, J., and Glauser, F.: Effect of carbon monoxide on exercise performance in chronic obstructive pulmonary disease. *Am J Med, 63*: 904-908, 1977.

Aronow, W.S., Stemmer, E.A., Wood, B., et al.: Carbon monoxide and ventricular fibrillation threshold in dogs with acute myocardial injury. *Am Heart J, 95*: 754-756, 1978.

Aronow, W.S., Stemmer, E.A., and Zweig, S.: Carbon monoxide and ventricular fibrillation threshold in normal dogs. *Arch Environ Health, 34*: 184-186, 1979.

Ashton, H., and Watson, D. W.: Puffing frequency and nicotine intake in cigarette smokers. *Br Med J, 3*: 679-681, 1970.

Ashton, H., Savage, R.D., Thompson, J.W., and Watson, D.W.: A method for measuring behavioural and physiological responses at different stress levels on a driving simulator. *Br J Pharmacol, 45*: 532-545, 1972.

Ashton, H., Savage, R.D., Telford, R., et al.: The effects of cigarette smoking on the response to stress in a driving simulator. *Br J Pharmacol, 45*: 546-556, 1972.

Ashton, H., and Telford, R.: Smoking and carboxyhaemoglobin. (Letter). *Lancet,* (ii): 857, 1973.

Asmussen, E., and Chiodi, H.: The effect of hypoxemia on ventilation and circulation in man. *Am J Physiol, 132*: 432-436, 1941.

Asmussmen, E., and Vinther-Paulsen, N.: On the circulatory adaptations to arterial hypoxemia (CO poisoning). *Acta Physiol Scand, 19*: 115-124, 1949.

Åstrup, P.: Carbon monoxide and peripheral vascular disease. *Scand J Clin Lab Invest, 19*: *Suppl 99*: pp. 193-197, 1967.

Åstrup, P.: Effects of hypoxia and of carbon monoxide exposures on experimental atherosclerosis. *Ann Intern Med, 71*: 426-427, 1969.

Åstrup, P.: Intraerythrocytic 2,3-diphosphoglycerate and carbon monoxide exposure. *Ann N Y Acad Sci, 174*: 252-254, 1970.

Åstrup, P.: Some physiological and pathological effects of moderate carbon monoxide exposure. *Br Med J, 4*: 447-452, 1972.

Åstrup, P.: Studies on carbon monoxide and nicotine. In Wynder, E.L., Hoffmann, D., and Gori, G.B. (Eds.): *Modifying the Risk for the Smoker. Vol. 1. Proceedings of the 3rd World Conference on Smoking and Health.* Washington, D.C., U.S. Govt. Print. Office, (DHEW Publication NIH 76-1221, 1976, pp. 331-341.

Åstrup, P., Hellung-Larsen, P., Kjeldsen, K., and Mellemgaard, K.: The effect of tobacco smoking on the dissociation curve of oxyhemoglobin. *Scand J Clin Lab Invest, 18*: 450-457, 1966.

Åstrup, P., Kjeldsen, K., and Wanstrup, J.: Enhancing influence of carbon monoxide on the development of atheromatosis in cholesterol-fed rabbits. *J Atheroscler Res, 7*: 343-354, 1967.

Åstrup, P., Kjeldsen, K., and Wanstrup, J.: Effects of carbon monoxide exposure on the arterial walls. *Ann NY Acad Sci, 174*: 294-300, 1970.

Åstrup, P., Olson, H.M., Trolle, D., and Kjeldsen, K. Effect of moderate carbon monoxide exposure on fetal development. *Lancet,* (ii): 1220-1222, 1972.

Ator, N.A., Merigan, W.H., and McIntire, R.W.: The effects of brief exposures to carbon monoxide on temporally differentiated responding. *Environ Res, 12*: 81-91, 1976.

Autian, J.: Medical aspects of toxicity resulting from fire exposure. In *Physiological and Toxicological Aspects of Combustion Products. NAS/NRC International Symposium,* Salt Lake City, 1976, pp. 47-56.

Ayres, S.M.: CO concentrations recorded on Ecolyzer and gas chromatograph. Personal communication, 1974.

Ayres, S.M., Giannelli, S., and Armstrong, R. G.: Carboxyhemoglobin: hemodynamic and respiratory responses to small concentrations. *Science, 149*: 193-194, 1965.

Ayres, S.M., Criscitiello, A. and Giannelli, S.: Determination of blood carbon monoxide content by gas chromatography. *J Appl Physiol, 21:* 1368-1370, 1966.

Ayres, S.M., Penny, J., Criscitiello, A., and Giannelli, S.: Effect of abnormal hemoglobin (carboxyhemoglobin) on human myocardial metabolism. *Clin Res, 16*: 220 (Abstr.), 1968.

Ayres, S.M., Mueller, H.S., Gregory, J.J., et al.: Systemic and myocardial hemodynamic responses to relatively small concentrations of carboxyhemoglobin (COHb). *Arch Environ Health, 18*: 699-709, 1969.

Ayres, S.M., Giannelli, S., and Mueller, H.: Myocardial and systemic responses to carboxyhemoglobin. *Ann NY Acad Sci, 174:* 268-293, 1970.

Ayres, S.M., Evans, R., Licht, D., et al.: Health effects of exposure to high concentrations of automotive emissions. *Arch Environ Health, 27*: 168-178, 1973.

Babcock, L.R., and Nagda, N.L.: Indices of air quality. In Thomas, W.A. (Ed.): *Indicators of Environmental Quality*. New York, Plenum Press, 1972, pp. 183-197.

Back, K.C.: *Effects of Carbon Monoxide on the Performance of Monkeys*. U.S. Aerospace Medical Research Laboratory Report AMRL-TR-69-130, Paper # 3, 1969.

Back, K.C., and Dominguez, A.M.: *Psychopharmacology of Carbon Monoxide under Ambient and Altitude Conditions*. U.S. Aerospace Medical Research Laboratory Report AMRL-TR-68-175, Paper, # 7, 1968.

Badre, R., Guillerm, R., Abran, N., et al.: Pollution atmosphérique par la fumée de tabac. *Ann Pharm Fr, 36*: 443-452, 1978.

Baker, S.P., Fisher, R.S., Masemore, W.C., and Sopher, I.M.: Fatal unintentional carbon monoxide poisoning in motor vehicles. *Am J Public Health, 62*: 1463-1467, 1972.

Baker, S.R., and Lilly, D.J.: Hearing loss from acute carbon monoxide intoxication. *Ann Otol Rhinol Laryngol, 86*: 323-328, 1977.

Balabaeva, L., and Kalpaznov, I.: Effect of smoking and length of service on the content of carboxyhemoglobin in blood of traffic policemen. *Higiena Zdraveopazavane, 5*: 490-493, 1974.

Ball, E.G., Strittmatter, C.F., and Cooper, O.: The reaction of cytochrome oxidase with carbon monoxide. *J Biol Chem, 193*: 635-647, 1951.

Barcroft, J.: *The Respiratory Function of the Blood*. London, Cambridge University Press, 1914.

Baretta, E.D., Stewart, R.D., Graff, S.A., and Donahoo, K.K.: Methods developed for the mass sampling analysis of CO and carboxyhaemoglobin in man. *Am Ind Hyg Assoc J, 39*: 202-209, 1978.

Barham, E.G., and Wilton, J.W.: Carbon monoxide production by a bathypelagic siphonophore. *Science, 144*: 860-862, 1964.

Barlow, D.H., and Baer, D.J.: Effect of cigarette smoking on the critical flicker frequency of heavy and light smokers. *Percept Mot Skills, 24*: 151-155, 1967.

Barnard, R.J., Gardner, G.W., and Diaco, N.V.: "Ischemic" heart disease in fire fighters with normal coronary arteries. *J Occup Med, 18*: 818-820, 1976.

Barois, A., Grosbuis, S., and Goulon, M.: Acute poisoning by carbon monoxide and heating gases. *Rev Prat, 29*: 1211-1214, 1979.

Barrowcliff, D.F.: Chronic carbon monoxide poisoning caused by methylene chloride paint-stripper. *Med Sci Law, 18*: 238, 1978.

Barrowcliff, D.F., and Knell, A.J.: Cerebral damage due to endogenous chronic carbon monoxide poisoning caused by exposure to methylene chloride. *J Soc Occup Med, 29*: 12-14, 1979.

Bartlett, O.: Patho-physiology of exposure to low concentrations of CO. *Arch Environ Health, 16*: 719-727, 1968.

Bates, D.R., and Witherspoon, A.E.: The photochemistry of some minor constituents of the earth's atmosphere. *R Astron Soc Monthly Notices, 112*: 101-124, 1952.

Bates, D.V. The uptake of carbon monoxide in health and in emphysema. *Clin Sci, 11*: 21-32, 1952.

Bauer, S., Cipeveski, V., and Grekovski, K.: Profesionalna Ekspozicija Ugljenmonoksidu u

Celicani Konvertorskog Tipa. *Arh Hig Rada Toksikol, 27*: 307-311, 1976.

Baumberger, J.P.: The carbon monoxide content of tobacco smoke and its absorption on inhalation. *J Pharmocol Exp Ther, 21*: 23-34, 1923.

Bay, H.W., Blurton, K.F., Lieb, H.L., and Oswin, H.G.: Electrochemical measurement of carbon monoxide. *Am Lab, 57*: 1972.

Bay, H.W., Blurton, K.F., Sedlak, J.M., and Valentine, A.M.: Electrochemical technique for the measurement of carbon monoxide. *Anal Chem, 46*: 1837-1839, 1974.

Beard, R.R.: Toxicological appraisal of carbon monoxide. *J Air Pollut Control Assoc, 19*: 722-727, 1969.

Beard, R.R., and Wertheim, G.A.: Behavioral impairment associated with small doses of carbon monoxide. *Am J Public Health, 57*: 2012-2022, 1967.

Beard, R.R., and Grandstaff, N.: Carbon monoxide exposure and cerebral function. *Ann NY Acad Sci, 174*: 385-395, 1970a.

Beard, R.R., and Grandstaff, N.: *Behavioral Responses to Small Doses of Carbon Monoxide.* U.S. Aerospace Medical Research Laboratory Report AMRL-TR-70-102, Paper 7, 1970b.

Beard, R.R., and Grandstaff, N.W.: *Carbon Monoxide and Human Functions.* Paper presented at Fifth Rochester International Conference on Environmental Toxicity: Behavioral Toxicology, June 7-9, 1972.

Beck, H.G., and Suter, G.M.: Role of carbon monoxide in the causation of myocardial disease. *JAMA, 110*: 1982-1988, 1938.

Beckman, A.O., McCullough, J.D., and Crane, R.A.: Microdetermination of carbon monoxide in the air. A portable instrument. *Anal Chem, 20*: 674-677, 1948.

Behrman, R.E., Fisher, D.E., and Patton, J.: Air pollution in nurseries. Correlation with a decrease in oxygen carrying capacity of hemoglobin. *J Pediatr, 78*: 66-74, 1971.

Bell, D.G., and Wright, G. R.: Energy expenditure and work stress of divers performing a variety of underwater tasks. *Ergonomics, 22*: 345-356, 1979.

Bell, M.A.: Sub-acute carbon monoxide poisoning. *Arch Environ Health, 3*: 594-596, 1961.

Bell, R.J.: *Introductory Fourier Transform Spectroscopy.* New York, Academic Press, 1972.

Bender, W., Gothert, M., Malorny, G., and Sebbesse, P.: Wirkungsbild niedriger Kohlenoxid-Konzentrationen beim Menschen. (Effects of low carbon monoxide concentrations in man). *Arch Toxikol, 27*:142-158, 1971.

Benesch, R., and Benesch, R.E.: The effect of organic phosphates from the human erythrocyte on the allosteric properties of hemoglobin. *Biochem Biophys Res Commun, 26*: 162-167, 1967.

Benesch, R.E., Maeda, N., and Benesch, R.: 2,3-Diphosphoglycerate and the relative affinity of adult and fetal hemoglobin for oxygen and carbon monoxide. *Biochim Biophys Acta (P), 257*: 178-182, 1972.

Benignus, V.A., and Otto, D.A.: Lack of effects of carbon monoxide on human vigilance. *Percept Mot Skills, 45*: 1007-1014, 1977.

Bensinger, T.A., Brouillard, R.P., and Conrad, M.E.: Endogenous production of carbon monoxide as measured simultaneously in blood and gas phase. *Br J Haematol, 29*: 487-494, 1975.

Benson, F.B., Henderson, J.J., and Caldwell, D.E.: *Indoor-Outdoor Air Pollution Relationships: A Literature Review.*Triangle Park, North Carolina, U.S. Environmental Protection Agency (Publication AP 112), 1972.

Benzon, H.T., Claybon, L, and Brunner, E.A.: Elevated carbon monoxide levels from exposure to methylene chloride. *JAMA, 239*: 2341, 1978.

Bergman, I., Coleman, J.E., and Evans, D.: Simple gas chromatograph with an electrochemical detector for the measurement of hydrogen and carbon monoxide in the parts per million range, applied to exhaled air. *Chromatographia, 8*: 581-583, 1975.

Bernard, M., Balny, C., Banerjee, R., and Douzon, F.: Recombination of carbon monoxide with hemoglobin after flash photolysis of the carboxy derivative in mixed solvents at sub zero temperature. *Biochim Biophys Acta, 393*: 389-395, 1975.

Bessouro, R., and Gray, J.: Carbon monoxide poisoning and nonoliguric acute renal failure. *Can Med Assoc J, 119*: 41-44, 1978.

Bharadway, H., Singh, A.P., and Malhotra, M.S.: Body composition of the high altitude natives of Ladakh. A comparison with sea level residents. *Hum Biol, 45*: 423-434, 1973.

Biersteker, K.: Air pollution control in the Netherlands. *Tijdschr Soc Geneesk, 50*: 17-21, 1972.

Biersteker, K.: Sulfur dioxide and suspended particulate matter. Where do we stand? *Environ Res, 11*: 287-304, 1976.

Birstingl, M., Hawkins, L., and McEwen, T.: Experimental atherosclerosis during chronic exposure to carbon monoxide. *Eur Surg Res, 2:* 92-93, 1970.

Bisgard, H.E., Will, J.A., Tyson, I.B., et al.: Distribution of regional lung function during mild exercise in residents of 3100 m. *Respir Physiol, 22*: 369-379, 1974.

Bissonnette, J.M., Wickham, W.K., and Drummond, W.H.: Placental diffusing capacities at varied carbon monoxide tensions. *J Clin Invest, 59*: 1038-1044, 1977.

Bithel, L.: *Report of Carbon Monoxide Assessment, Toronto International Airport.* Toronto, Occupational Health Services, Ontario Department of Health, July 17, 1970.

Blackmore, D.J.: The determination of carbon monoxide in blood and tissue. *Analyst, 95*: 439-458, 1970.

Blackmore, D.J.: Aircraft accident toxicology: U.K. experience 1967-1972. *Aerospace Med, 45*: 987-994, 1974.

Bloomfield, B.D.: Lift trucks raise carbon monoxide level. *Arch Ind Health, 15*: 172-173, 1957.

Bloomfield, J.J., and Isbell, H.S.: The problem of automobile exhaust gas in streets and repair shops of large cities. *Public Health Rep, 43*: 750-765, 1928.

Blurton, K.F., and Bay, H.W.: Controlled-potential electrochemical analysis of carbon monoxide. *Am Lab, 6*: 50-52, 54-55, 1974.

Boatman, J.B., and Carter, S.D.: Hypoxia and the arterial surface. *Arch Environ Health, 17*: 360-363, 1973.

Bogen, E.: The composition of cigarettes and cigarette smoke. *JAMA, 93*: 1110-1114, 1929.

Bokhaven, C., and Niessen, H.J.: Amounts of oxides of nitrogen and carbon monoxide in cigarette smoke, with and without inhalation. *Nature, 192*: 458-459, 1961.

Bondi, K.R., Very, K.R., and Schaefer, K.E.: Carboxyhemoglobin levels during a submarine patrol. *Aviat Space Environ Med, 49*: 851-854, 1978a.

Bondi, K.R., Very, K.R., and Schaefer, K.E.: Carboxyhemoglobin levels during a submarine patrol. *Undersea Biomed Res, 5*: 17A, 1978b.

Bondi, K.R., and Schaefer, K.E.: Carboxyhemoglobin (COHb) in smokers and non-smokers and ambient levels of nitrogen dioxide (NO_2) as a result of smoking during long patrols. *Toxicology Research Projects Directory, 4 (10)*, 1979.

Bonham, G.H.: Fasten your seat belts—no smoking. *Br Columbia Med J, 15*: 304-305, 1973.

Boudene, C., Belegaud, J., Jouany, J.M., and Truhaut, R.: In-vivo carbon monoxide production after dichloromethane exposure. Experimental inhalation study on rabbits. Contribution of these data to an eventual modification of the threshold limit value of this solvent. *Arch Mal Prof Med Trav Secur Soc, 39*: 657-669, 1978.

Bove, J.L., and Siebenberg, S.: Airborne lead and carbon monoxide at 45th Street, New York City. *Science, 1967*: 986-987, 1970.

Bowditch, F.W.: *Progress and Future of Automotive Emission Control–U.S. and Canada.* Paper T.A.2. Presented at International Conference on Automobile Pollution, Toronto, Canada, June 26-28, 1972.

Bowne, N., Boyer, A.E., Trent, K.E., and Cooper, D.G.: *An Air Quality Model for Metropolitan Toronto.* Paper presented at 64th Annual Meeting of the Air Pollution Control Association, Atlantic City, New Jersey, June 27-July 1, 1971.

Brewster, E., Collins, S., Funnell, G.R., and Smith, E.B.: The effect of high pressure on the hemolysis of red blood cells. *Undersea Biomed Res, 3*: 151-155, 1976.

Breysse, P.A.: Chronic carbon monoxide poisoning. *Ind Med Surg, 30*: 20-22, 1961.

Breysse, P.A., and Bovee, H.H.: Use of expired air-carbon monoxide for carboxyhemoglobin determinations in evaluating carbon monoxide exposures resulting from the operation of gasoline fork lift trucks in holds of ships. *Am Ind Hyg Assoc J, 30*: 477-483, 1969.

Brice, R.M., and Roesler, J.F.: The exposure to carbon monoxide of occupants of vehicles moving in heavy traffic. *J Air Pollut Control Assoc, 16*: 597-600, 1966.

Bridge, D.P., and Corn, M.: Contribution to the assessment of exposure of non-smokers to air pollution from cigarette and cigar smoke in occupied spaces. *Environ Res, 5*: 192-209, 1972.

Brief, R.S., Jones, A.R., and Yoder, J.D.: Lead, carbon monoxide and traffic: a correlation study. *J Air Pollut Control Assoc, 10*: 384-388, 1960.

Brief, R.S., Ajemina, R.S., and Confer, R. G.: Iron pentacarbonyl: its toxicity, detection and potential for formation. *Am Ind Hyg Assoc J, 28*: 21-30, 1967.

Brieger, H.: Carbon monoxide polycythemia. *J Ind Hyg Toxicol, 26*: 321-327, 1944.

Brinkhouse, M.M.: *Effects of Low Level Carbon Monoxide Exposure, Blood Lipids and Coagulation Parameters.* Triangle Park, North Carolina, U.S. Environmental Protection Agency (Health Effects Research Lab Tech. Inf. Service PB - 269-340), 1977, pp. 1-24.

British Medical Journal: Breathing other people's smoke. *Br Med J, 2:* (6135) 453-454, 1978.

British Standards Institute: *Respirators for Protection against Harmful Dusts, Gases, and Scheduled Agricultural Chemicals.* BS 2091, 1969.

Brody, J.S., and Coburn, R.F.: Carbon monoxide—induced arterial hypoxemia. *Science, 164*: 1297-1298, 1969.

Brody, J.S., and Coburn, R.F.: Effects of elevated carboxyhemoglobin on gas exchange in the lung. *Ann NY Acad Sci, 174*: 255-260, 1970.

Brown, I.D., and Poulton, E.C.: Measuring the spare "mental capacity" of car drivers by a subsidiary task. *Ergonomics, 4*: 35-40, 1961.

Buchwald, H.: A rapid and sensitive method for estimating carbon monoxide in blood and its application in problem areas. *Am Ind Hyg Assoc J, 30*: 564-569, 1969a.

Buchwald, H.: Exposure of garage and service station operatives to carbon monoxide: a survey based on carboxyhemoglobin levels. *Am Ind Hyg Assoc J, 30*: 570-575, 1969b.

Burch, D.E., Gates, F.J., and Pembrook, J.D.: Ambient CO monitor. Triangle Park, North Carolina, United States Environmental Protection Agency (Tech Rpt. EPA 660/2-76-210), 1976.

Burgess, W.A., Diberadinis, L., and Speizer, F.E.: Health effects of exposure to automobile exhaust. Part 5. Exposure of toll booth operators to automobile exhaust. *Am Ind Hyg Assoc J, 38*: 184-191, 1977.

Burgess, W.A., Sidor, R., Lynch, J.J., et al.: Minimum protection factors for respiratory protective devices for fire fighters. *Am Ind Hyg Assoc J, 38:* 18-32, 1977.

Burns, D.M.: Consequences of smoking—the involuntary smoker. In Steinfeld, J., Griffiths, W., Ball, K., and Taylor, R.M. (Eds.): *Health Consequences, Education, Cessation Activities, and Governmental Action. Vol. II. Proceedings of the 3rd World Conference on Smoking and Health.* Washington, D.C., U.S. Govt. Print. Office, (DHEW Publication 77-1413), 1977, pp. 51-58.

Burri, P.H., and Weibel, E.R.: Morphometric evaluation of changes in lung structure due to high altitude. In Porter, R., and Knight, J. (Eds.): *High Altitude Physiology: Cardiac and Respiratory Aspects.* Edinburgh: Churchill-Livingstone, 1971.

Busey, W.M.: *Summary Report: Study of Synergistic Effects of Certain Airborne Systems in the Cynomolgus Monkey.* Vienna, Virginia, Hazleton Laboratories, Inc., 1972.

Buskirk, E.R., Kollias, J., Piconreatigue, E., et al.: Physiology and performance of track athletes at various altitudes in the United States and Peru. In Goddard, R.F., (Ed.): *International Symposium on the Effects of Altitude on Physical Performance.* Chicago, Athletic Institute, 1967.

Butler, N.R., and Goldstein, H.: Smoking in pregnancy and subsequent child development.

Br Med J, 4: 573-575, 1973.

Butt, I., Davies, G.M., Jones, J.G., and Sinclair, A.: Carboxyhemoglobin levels in blast furnace workers. *Ann Occup Hyg, 17*: 57-63, 1974.

Byrne-Quinn, E., Weil, J.W., Sodal, I.E., et al.: Ventilatory control in the athlete. *J Appl Physiol, 30*: 91-98, 1971.

California Transportation Agency: *The Roles of Carbon Monoxide, Alcohol and Drugs in Fatal Single Car Accidents.* California Transportation Agency, 1965.

Calverley, P.M.A., McElderry, L., Leggett, R.J.E., and Flenley, D.C.: Carbon monoxide, oxygen transport and exercise tolerance in chronic bronchitis and emphysema. *Clin Sci 58:* 5P, 1979.

Cameron, B.F., Lian, C.Y., Carravazalino, J., et al.: Regulation of oxygen dissociation by 2:3 diphosphoglycerate in the human erythrocyte. In *Molecular Basis of Biological Activity. PAABS Symposium Vol. 1.* New York, Academic Press, 1972, pp. 169-196.

Campbell, J.A.: Tissue oxygen tension and carbon monoxide poisoning. *J Physiol, 68*: 81-96, 1929.

Cano, J.P., Catalin, J., Badré, R., et al.: Determinations of nicotine by gas-phase chromatography II. Applications. *Ann Pharm Fr, 28*: 633-640, 1970.

Carlsson, A., and Huttengren, M.: Exposure to methylene chloride. Part 3. Metabolism of carbon-14 labelled methylene chloride in rat. *Scand J Work Environ Health, 1*: 104-108, 1975.

Carlsten, A., Holmgren, A., Linroth, K., et al.: Relationship between low values of alveolar carbon monoxide concentration and carboxyhemoglobin percentage in human blood. *Acta Physiol Scand, 31:* 62-74, 1954.

Carnelley, T., and Haldane, J.S.: The air of sewers. *Proc R Soc Lond, 42*: 501-522, 1887.

Case, D.A., and Karplus, M.: Stereochemistry of carbon monoxide binding to myoglobin and hemoglobin. *J Mol Biol, 123:* 697-702, 1978.

Castleden, C.M., and Cole, P.V.: Inhalation of tobacco smoke by pipe and cigar smokers. *Lancet (ii):* 21-22, 1973.

Chaney, L.W.: Carbon monoxide automobile emissions measured from the interior of a travelling automobile. *Science, 199:* 1203-1204, 1978.

Chanutin, A., and Curnish, R.R.: Effect of organic and inorganic phosphates on the oxygen equilibrium of human erythrocytes. *Arch Biochem Biophys, 121*: 96-102, 1967.

Chapman, D.J., and Tocher, R.D.: Occurrence and production of carbon monoxide in some brown algae. *Can J Botany, 44:* 1438-1442, 1966.

Chappell, S.B., and Parker, R.J.: Smoking and carbon monoxide levels in enclosed public places in New Brunswick. *Can J Public Health, 68:* 159-161, 1977.

Chass, R.L., Hamming, W.J., Dickinson, J.E., and MacBeth, W.G.: *Los Angeles Photo-Chemical Smog: Past, Present and Future.* Paper M.A. 5. presented at International Conference on Automobile Pollution, Toronto, Canada, June 26-28, 1972.

Chaussin, M., and Guenard, H.: Diffusion limitation in pulmonary gas exchange. *Bull Eur Physiopathol Respir, 13:* 153P-145P, 1977.

Cheetham, R.W.S.: *Report on Tests Carried out on Aircraft for CO Contamination. Report 16, Psycho-Physiological Laboratory.* Pretoria, Republic of South Africa, South African Airforce, 1944.

Chevalier, R.B., Krumholtz, R.A., and Ross, J.C.: Reaction of non-smokers to carbon monoxide inhalation, cardiopulmonary responses at rest and during exercise. *JAMA, 198:* 1061-1064, 1966.

Chiodi, H., Dill, D.B., Consolazio, F., and Horvath, S.M.: Respiratory and circulatory responses to acute carbon monoxide poisoning. *Am J Physiol, 134:* 683-693, 1941.

Chipman, J.C., and Massey, M.T.: Proportional sampling system for the collection of an integrated auto exhaust gas sample. *J Air Pollut Control Assoc, 10:* 60-69, 1960.

Chovin, P.: Carbon monoxide: analysis of exhaust gas investigations in Paris. *Environ Res, 1:* 198-216, 1967.

Chovin, P. Discussion. In Berlin, A., and Smeets, J. (Eds.): *Proceedings. Health Effects of Carbon Monoxide Environmental Pollution.* Luxembourg, Commission of the European Communities Health Protection Directorate, 1974a, pp. 225-236.

Chovin, P.: Le monoxyde de carbone. In Coulston, F., and Korte, F. (Eds.): *Environmental Quality and Safety. Global Aspects of Chemistry, Toxicology and Technology as Applied to the Environment, Vol. 3.* Stuttgart, George Thieme Publishers/New York, Academic Press, 1974b.

Chovin, P., and Truffert, L.: Atmospheric carbon monoxide pollution in urban thoroughfares and road tunnels in the EEC countries. In *European Colloquium: Health Effects of Carbon Monoxide Environmental Pollution.* Luxembourg, Directorate General of Social Affairs—Health Protection Directorate EEC, 1973, pp. 115-120.

Christensen, C.L., Gliner, J.A., Horvath, S.M., and Wagner, J.A.: Effects of three kinds of hypoxias on vigilance performance. *Aviat Space Environ Med, 48:* 491-496, 1977.

Christianson, R.E.: Gross differences observed in the placentas of smokers and nonsmokers. *Am J Epidemiol, 110:* 178-187, 1979.

Christman, A.A., and Randall, E.L.: A convenient and accurate method for the determination and detection of carbon monoxide in blood. *J Biol Chem, 102:* 595-609, 1933.

Ciampolini, E.: Carbon monoxide hazard in public garages. *J Ind Hyg, 6:* 102-109, 1924.

Clark, B.J., and Coburn, R.F.: Mean myoglobin oxygen tension during exercise at maximal oxygen uptake. *J Appl Physiol, 39:* 135-144, 1975.

Clark, D.W., and MacMahon, B.: *Preventive Community Medicine* (2nd ed.). Boston, Little, Brown, 1981.

Clark, R.T., and Otis, A.B.: Comparative studies on acclimatization of mice to carbon monoxide and to low oxygen. *Am J Physiol, 196:* 285-294, 1952.

Claude, J.R., Lellouch, J., and Truhaut, R.: Epidemiological study of the variations of carboxyhaemoglobinemia with smoking habits in 1,414 male human subjects. Demonstration of the permanence of carbon monoxide in blood. *Arch Mal Prof Med Trav Secur Soc, 39:* 333-337, 1978.

Claude, J.R., Lellouch, J., and Thevinin, M.: Air pollution and carboxyhaemoglobin in man. In Benarie, M. (Ed.): *Studies in Environmental Science I.* Amsterdam, Elsevier Scientific Publishing, 1978, pp. 113-115.

Clayton, G.D., Cook, W.A., and Frederick, W.G.: A study of the relationship of street level carbon monoxide concentrations to traffic accidents. *Am Ind Hyg Assoc J, 21:* 46-54, 1960.

Coburn, R.F.: Endogenous carbon monoxide production and body CO stores. *Acta Med Scand Suppl, 472:* 269-282, 1967.

Coburn, R.F.: The carbon monoxide body stores. *Ann NY Acad Sci, 174:* 11-22, 1970.

Coburn, R.F.: Endogenous carbon monoxide metabolism. *Ann Rev Med, 24:* 241-250, 1973.

Coburn, R.F. (Ed.): *The Biologic Effects of Carbon Monoxide on the Human Organism. Report of a Committee of the National Research Council—National Academy of Sciences.* Washington, D.C., U.S. Govt. Print. Office, 1977.

Coburn, R.F., Blakemore, W.S., and Forster, R.E.: Endogenous carbon monoxide production in man. *J Clin Invest, 42:* 1172-1178, 1963.

Coburn, R.F., Danielson, G.K., Blakemore, W.S., and Forster, R.E.: Carbon monoxide in blood: analytical method and sources of error. *J Appl Physiol, 19:* 510-519, 1964.

Coburn, R.F., Williams, W.J., and Forster, R.E.: Effect of erythrocyte destruction on carbon monoxide production in man. *J Clin Invest, 43:* 1098-1103, 1964.

Coburn, R.F., Forster, R.E., and Kane, P.B.: Considerations of the physiological variables that determine the blood COHb concentrations in man. *J Clin Invest, 44:* 1899-1910, 1965.

Coburn, R.F., Blakemore, W.S., and Forster, R.E.: Endogenous carbon monoxide production in patients with hemolytic anaemia. *J Clin Invest, 45:* 460-468, 1966.

Coburn, R.F., Williams, W.J., White, P., and Kahn, S.B.: The production of carbon monoxide from hemoglobin in vivo. *J Clin Invest, 46:* 346-356, 1967.

Coburn, R.F., Ploegmakers, F., Gondrie, P., and Abboud, R.: Myocardial myoglobin oxygen tension. *Am J Physiol, 224:* 870-876, 1973.

Cohen, S.I., Deane, M., and Goldsmith, J.R.: Carbon monoxide and survival from myocardial infarction. *Arch Environ Health, 19:* 510-517, 1969.

Cohen, S.I., Dorion, G., Goldsmith, J.R., and Permutt, S.: Carbon monoxide uptake by inspectors at a United States-Mexico Border Station. *Arch Environ Health, 22:* 47-54, 1971.

Cohen, S.I., Perkins, N.M., Ury, H.K., and Goldsmith, J.R.: Carbon monoxide uptake in cigarette smoking. *Arch Environ Health, 22:* 55-60, 1971.

Cole, P.: Carbon monoxide, environmental pollution and public health. Luxembourg, 17th-19th December, 1973. Cited by Commins (1975).

Cole, P.V.: Uptake of CO by man: a population study of COHb levels. In Berlin, A., and Smeets, J. (Eds.): *Proceedings Health Effects of Carbon Monoxide Environmental Pollution.* Luxembourg, Commission of European Communities Health Protection Directorate, 1974, pp. 151-167.

Cole, P.V., Hawkins, L.H., and Roberts, D.: Smoking during pregnancy and its effects on the fetus. *J Obstet Gynaecol Br Commonw, 79:* 782-787, 1972.

Collier, C.R.: Oxygen affinity of human blood in presence of carbon monoxide. *J Appl Physiol, 40:* 487-490, 1976.

Collison, H.A., Rodkey, F.L., and O'Neal, J.D.: Determination of carbon monoxide in blood by gas chromatography. *Clin Chem, 14:* 162-171, 1968.

Colucci, A.V., Hammer, D.I., Williams, M.E., et al.: Pollutant burdens and biological response. *Arch Environ Health, 27:* 151-154, 1973.

Colucci, J.M., and Begeman, C.R.: Carbon monoxide in Detroit, New York and Los Angeles air. *Environ Sci Technol, 3:* 41-47, 1969.

Commins, B.T.: Measurement of carbon monoxide in the blood: review of available methods. *Ann Occup Hyg, 18:* 69-77, 1975.

Commins, B.T., and Lawther, P.J.: A sensitive method for the determination of carboxyhaemoglobin in a finger prick sample of blood. *Br J Ind Med, 22:* 139-143, 1965.

Comroe, J.H.: *Physiology of Respiration.* Chicago, Year Book Medical Publishers, 1965.

Comroe, J.H., Forster, R.E., DuBois, A.B., et al.: *The Lung.* Chicago, Year Book Medical Publishers, 1970.

Conlee, C.J., Kenline, P.A., Cummins, R.L., and Konopinski, V.J.: Motor vehicle exhaust at three selected sites. *Arch Environ Health, 14:* 429-446, 1967.

Conway, E.J.: *Microdiffusion Analysis and Volumetric Error,* 3rd ed. London, Crosby, Lockwood & Son, 1950, pp. 1-391.

Cooper, A.G.: *Carbon Monoxide: A Bibliography with Abstracts.* Washington, D.C., U.S. Govt. Print. Office (DHEW Public Health Service Publication 1503), 1966.

Cooper, D.Y., Schleyer, H., Rosenthal, O., et al.: Inhibition by CO of hepatic benzo α pyrene hydroxylation and its reversal by monochromatic light. *Eur J Biochem, 74:* 69-75, 1977.

Corcoran, D.W.J.: Personality and the inverted-U relation. *Br J Psychol, 52:* 267-273, 1965.

Corso, J.F.: *The Experimental Psychology of Sensory Behavior.* New York, Holt, Rinehart and Winston, Inc., 1967.

Corya, B.C., Black, M.J., and McHenry, P.L.: Echocardiographic findings after acute carbon monoxide poisoning. *Br Heart J, 38:* 712-717, 1976.

Cote, W.A., et al.: *A Study of Indoor Air Quality: Final Report (EPA Contract 68-02-0745).* Triangle Park, North Carolina, U.S. Environmental Protection Agency, 1974.

Council on Environmental Quality. *2nd Annual Report.* Washington, D.C., U.S. Govt. Print. Office, 1971.

Council on Environmental Quality. *Environmental Quality—the Eighth Annual Report of the*

Council on Environmental Quality, 1977. Washington, D.C., U.S. Govt. Print. Office (1978-0-237-689), 1978.

Creighton, D.E., and Lewis, P.H.: The effect of different cigarettes on human smoking patterns. In Thornton, R.E. (Ed.): *Smoking Behaviour. Physiological and Psychological Influences.* Edinburgh, Churchill-Livingstone, 1978, pp. 289-300.

Cropp, G.J.A.: Ventilation in anemia and polycythemia. *Can J Physiol Pharmacol, 48:* 382-393, 1970.

Croton, L.M.: Carbon monoxide poisoning: A current controversy. (Abstr.). *Arch Environ Health, 1:* 149, 1959.

Crozier, W.J., and Holway, A.H.: Theory and measurement of visual mechanisms. I. A visual discriminometer. *J Gen Physiol, 22:* 341-351, 1939.

Crutzen, P.J., and Fishman, J.: Average concentrations of OH in the troposphere, and the budgets of CH_4, CO, H_2 and $CH_3C\ Cl_3$. *Geophys Res Ltr, 4:* 321-324, 1977.

Cucu, F.: Carbon monoxide and its implications in atherosclerosis etiopathogeny. *Med Interne, 16:* 229-242, 1978.

Cuddeback, J.E., Donovan, J.R., and Burg, W.R.: Occupational aspects of passive smoking. *Am Ind Hyg Assoc J, 37:* 263-267, 1976.

Curphey, T.J., Hood, L.P.L., and Perkins, N.M.: Carboxyhemoglobin in relation to air pollution and smoking. *Arch Environ Health, 10:* 179-185, 1965.

Dahlström, H.: A critical study of Sjöstrand's method for determination of COHb and the total hemoglobin of the human body. *Acta Physiol Scand, 33:* 296-308, 1955.

Dahms, T.E., and Horvath, S.M.: Rapid, accurate technique for determination of carbon monoxide in blood. *Clin Chem, 20:* 533-537, 1974.

Dale, B.T.: Carbon monoxide asphyxiation from use of a charcoal barbecue in homemade wooden trailer. Personal communication, 1973.

Dalhamn, T., Edfors, M.L., and Rylander, R.: Retention of cigarette smoke components in human lungs. *Arch Environ Health, 17:* 746-748, 1968.

Daly, M. de B, Lambertsen, C.J., and Schweitzer, A.: Observations on the volume of blood flow and oxygen utilization of the carotid body in the cat. *J Physiol, 125:* 67-89, 1954.

Davidson, S.: *The Principles and Practice of Medicine,* 9th ed. Edinburgh, E. and S. Livingstone Ltd., 1970.

Davie, R., Butler, N., and Goldstein, H.: *From Birth to Seven. The Second Report of the National Child Development Study (1958 Cohort),* London, Longman, 1972, pp. 1-198.

Davies, D.M.: The application of threshold limit values for carbon monoxide under conditions of continuous exposure. *Ann Occup Hyg, 18:* 21-28, 1975.

Davies, D.M., and Smith, D.J.: Electrocardiographic changes in healthy men during continuous low-level carbon monoxide exposure. *Environ Res, 21:* 197-206, 1980.

Davies, D.P., Gray, O.P., Ellwood, P.C., and Abernethy, M.: Cigarette smoking in pregnancy: associations with maternal weight gain and fetal growth. *Lancet,* (i): 385-387, 1976.

Davies, G.M., Jones, J.G., and Warner, C.G.: A continuously recording atmospheric carbon monoxide monitoring system with fully automatic alarms in a blast furnace area. *Br J Ind Med, 22:* 270-278, 1965.

Davies, R.F., Topping, D.L., and Turner, D.M.: The effect of intermittent carbon monoxide exposure on experimental atherosclerosis in the rabbit. *Atherosclerosis, 24:* 527-536, 1976.

Deacon, J.: *Tobacco Tortured.* London, 1616. Published in facsimile by Da Capo Press, Amsterdam, 1968.

DeBias, D.A., Banerjee, C.M., Birkhead, N.C., et al.: Carbon monoxide inhalation effects following myocardial infarction in monkeys. *Arch Environ Health, 27:* 161-167, 1973.

DeBias, D.A., Banerjee, C.M., Birkhead, N.C., et al.: Effects of carbon monoxide inhalation on ventricular fibrillation. *Arch Environ Health, 31:* 42-46, 1976.

De Bruin, A.: Carboxyhemoglobin levels due to traffic exhaust. *Arch Environ Health, 15:* 384-389, 1967.

Delarue, N.C.: A study in smoking withdrawal. *Can J Public Health, 64:* S5-S19, 1973.

Delivoria-Papadopoulos, M.: The pregnant woman and the fetus. In DuBois, A.B. (Chairman): *Effects of Carbon Monoxide on Man's Health in his Home Environment.* Washington, D.C., U.S. Govt. Print. Office (DHEW Publication), 1973.

Delivoria-Papadopoulos, M., Coburn, R.F., and Forster, R.E.: Clinical variation of heme destruction and carbon monoxide production ($\dot{V}_{co}$) in the normal female. *Physiologist, 13:* 178, 1970.

Delivoria-Papadopoulos, M., Coburn, R.F., and Forster, R.E.: Cyclic variation of rate of carbon monoxide production in normal women. *J Appl Physiol, 36:* 49-51, 1974.

Dempsey, J.A.: Quantity and significance of pulmonary adaptations to work in adolescent altitude residents. *Acta Paediatr Scand Suppl, 217:* 99-104, 1971.

Denson, R., Nanson, J.L., and MacWatters, M.A.: Hyperkinesis and maternal smoking. *Can Psychiatr Assoc J, 20:* 183-187, 1975.

de Quiros, J.F.B., and Hess, M.H.H.: El fumar como factor de riesgo de las angiopatias obliterantes: investigaciones sobre los cambios iniciales en los vasos. *Rev Clin Esp, 143:* 351-356, 1976.

Derouanne, A., and Verduyn, G.: Etude de quelques facteurs influençant la pollution de l'air a l'intérieur des bâtiments. *Tribune du Cebedeau, 27:* 282-288, 1974.

Diamond, P.: Industrial hygiene aspects of carbon monoxide. McClelland Air Force Base, California, U.S. Air Force Environmental Health (Lab Report AD 784845), 1974, pp. 1-30.

Dill, D.B.: *Handbook of Physiology—Section 4. Adaptation to the Environment.* Washington, D.C., American Physiological Society, 1964, pp. 1-1056.

Dinman, B.D.: Pathophysiologic determinants of community air quality standards for carbon monoxide. *J Occup Med, 10:* 446-456, 1968.

Dinman, B.D.: Effects of long-term exposure to carbon monoxide. In DuBois, A.B. (Ed.): *Effects of Chronic Exposure to Low Levels of Carbon Monoxide on Human Health, Behavior and Performance.* Washington, D.C., National Academy of Sciences and National Academy of Engineering, 1969, pp. 25-38.

Dinman, B.D.: The management of acute carbon monoxide intoxication. *J Occup Med, 16:* 662-664, 1974.

Dinman, B.D., Eaton, J.W., and Brewer, G.J.: Effects of carbon monoxide on DPG concentrations in the erythrocyte. In Coburn, R.F. (Ed.): Biological effects of carbon monoxide. *Ann NY Acad Sci, 174:* 246-251, 1970.

Dixon, F.R.S.: The drug habit. *Br Med J, 2:* 819-822, 1921.

Dominguez, A.M.: Symposium—fire and incendiarism: problems of carbon monoxide in fires. *J Forensic Sci, 7:* 379-390, 1962.

Donald, K.W., and Paton, W.O.M.: Gases administered in artificial respiration with particular reference to the use of carbon monoxide. *Br Med J, 1:* 313-318, 1955.

Donham, K.J., Rubino, M., Thedell, T.D., and Kammermeyer, J.: Potential health hazards to agricultural workers in swine confinement buildings. *J Occup Med, 19:* 383-387, 1977.

Dorcus, R.M., and Weigand, G.E.: The effect of exhaust gas on the performance in certain psychological tests. *J Gen Psychol, 2:* 73-96, 1929.

Douglas, C.G.: John Scott Haldane, 1860-1936. In Cunningham, D.J.C., and Lloyd, B.B. (Eds.): *The Regulation of Human Respiration.* Oxford, Blackwell Scientific, 1963, pp. 3-32.

Douglas, C.G., Haldane, J.S., and Haldane, J.B.S.: The laws of combination of haemoglobin with carbon monoxide and oxygen. *J Physiol, 44:* 275-304, 1912.

Douglas, T.A., Lawson, D.D., Ledingham, I., et al.: Carbogen in experimental carbon monoxide poisoning. *Br Med J, 2:* 1673-1675, 1961.

Douglas, T.A., Ledingham, I., Lawson, D.D., and Norman, J.N.: Carbon monoxide poisoning. A comparison between the efficiencies of oxygen at one atmosphere pressure, of oxygen at two atmospheres pressure and of 5% and 7% carbon dioxide in oxygen. *Lancet,* (i): 68-69, 1962.

Douze, J.M.C.: Carbon monoxide poisoning is not obsolete. A form of cerebral oedema in internal medicine. *Neth J Med, 20:* 167-161, 1977.

Draper, A.B., Choi, H., and Robison, J.W.: Supplementary air to minimize carbon monoxide in cupola effluent. *Trans Am Foundrymen's Soc, 86:* 487-494, 1978.

Drinker, C.K.: *Carbon Monoxide Asphyxia.* New York, Oxford University Press, 1938.

Drinkwater, B.L., Raven, P.B., Horvath, S.M., et al.: Effect of carbon monoxide and peroxy-acetyl nitrate on man's ability to work in the heat. *Arch Environ Health, 28:* 177-181, 1974.

Dublin, L.T., and Vane, R.J.: Occupational mortality experience of insured wage earners. *Month Labor Rev, 64:* 1003, 1947.

DuBois, A.B. (Chairman): *Effects of Carbon Monoxide on Man's Health in his Home Environment.* Washington, D.C., U.S. Govt. Print. Office (DHEW Publication), 1973.

DuBois, A.B. (Ed.): *Effects of Chronic Exposure to Low Levels of Carbon Monoxide on Human Health, Behavior, and Performance.* Washington, D.C., National Academy of Sciences and National Academy of Engineering, 1969.

DuBois, L., Zdrojewski, A., and Monkman, J.L.: The analysis of carbon monoxide in urban air at the ppm level, and the normal carbon monoxide value. *J Air Pollut Control Assoc, 16:* 135-139, 1966.

Duck, B.W.: *Proceedings of the Symposium on the Assessment of Exposure and Risk.* Society of Occupational Medicine, 1973. Cited by Morrow (1975).

Dudley, W.L.: The poisonous effects of cigarette smoking. *Med News, 53:* 286-288, 1888.

Duke, H.N., Green, J.H., and Neil, E.: Carotid chemoreceptor impulse activity during inhalation of CO mixtures. *J Physiol, 118:* 520-527, 1952.

Durnin, J.V.G.A., and Passmore, R.: *Energy, Work and Leisure.* London, Heinemann, 1967.

Dyer, R.S., Eccles, C.V., Schwartzwelder, H.S., et al.: Prenatal carbon monoxide and adult evoked potentials in rat. *J Environ Sci Health,* C13: 107-120, 1979.

Eagan, C.J.: Introduction and terminology: habituation and peripheral tissue adaptations. *Fed Proc, 22:* 930-933, 1963.

Eberle, A.: Rechtliche Aspekte und Akzente des Passivrauchens. Interdisciplinare Dekanstrosse. Teil 2. Die Aktivitat der Passiv-raucher—ihr Einstieg in die Forensische Praxis. *Therapie der Gegenwart, 116:* 1984-2000, 1977.

Eckardt, R.E., MacFarland, H.N., Alarie, Y.C.E., and Busey, W.M.: The biologic effect from long term exposure of primates to carbon monoxide. *Arch Environ Health, 25:* 381-387, 1972.

Edwards, C.W.: The carotid body in animals at high altitude. In Porter, R., and Knight, J. (Eds.): *High Altitude Physiology: Cardiac and Respiratory Aspects.* Edinburgh, Churchill-Livingstone, 1971.

Ehrich, W.E., Bellet, W., and Lewey, F. H.: Cardiac changes from CO poisoning. *Am J Med Sci, 208*: 511-523, 1944.

Eisenbud, M., and Ehrlich, L.R.: Carbon monoxide concentration trends in urban atmospheres. *Science, 176*: 193-194, 1972.

Ejam-Berdyev, A.K.: Effect of small CO concentrations on the cardiovascular system of workers engaged in cast-iron moulding. *Harkovaskovo Med Inst, 110*: 185-186, 1973.

Ekblom, B., and Huot, R.: Response to submaximal and maximal exercise at different levels of carboxyhaemoglobin. *Acta Physiol Scand, 86*: 474-482, 1972.

Elfimova, E.V., Gusev, M.I., Nokikov, J.V., et al.: Study of the combined-systemic effect of atmospheric pollutants (gases and dusts). *Gig Sanit, 8*: 11-15, 1972.

Elliott, M.A., Nebel, G.J., and Rounds, F.G.: The composition of exhaust gases from diesel, gasoline and propane-powered motor coaches. *J Air Pollut Control Assoc, 5*: 103-108, 1955.

Engel, R.R., Rodkey, F.L., O'Neil, J.D., and Collison, H.A.: Relative affinity of human fetal hemoglobin for carbon monoxide and oxygen. *Blood, 33*: 37-45, 1969.

Engel, R.R., Rodkey, F.L., and Krill, C.E.: Carboxyhemoglobin as an index of hemolysis. *Pediatrics, 47*: 723-730, 1971.

Environment Agency: *Quality of the Environment in Japan.* Tokyo, Environment Agency, 1975.

Environmental Criteria and Assessment Office: *Preliminary Assessment of Adverse Health Effects from Carbon Monoxide and Implications for Possible Modification of the Standard.* Triangle Park, North Carolina, United States Environmental Protection Agency, June 1979.

Environmental Protection Agency: *National Inventory of Air Pollutant Emissions, 1969.* Washington, D.C., U.S. Government Printing Office, 1969.

Environmental Protection Agency: National primary and secondary ambient air quality standards. *Federal Register, 36 (84)*: 8186-8201, 1971a.

Environmental Protection Agency: *Implementation Plans. Federal Register 36* (158): 15486-15506, 1971B.

Environmental Protection Agency: *Nationwide Emissions Report.* Triangle Park, North Carolina U.S. Environmental Protection Agency, 1973.

Environmental Protection Agency: Ambient air monitoring reference and equivalent methods. *Federal Register, 40*: 7042-7070, 1975.

Environmental Protection Agency: *National Air Quality and Emissions Trends Report 1977.* Triangle Park, North Carolina, U.S. Environmental Protection Agency (EPA 450/78/000), 1978.

Environmental Protection Agency: *National Air Quality, Monitoring and Emissions Trend Report, 1977.* Triangle Park, North Carolina, U.S. Environmental Protection Agency (EPA-450/2-78-52), December, 1978.

Environment Canada: *National Air Quality Objectives.* News Release, January 3, 1973.

Epstein, G.H.: Coronary heart disease epidemiology revisited: clinical and community aspects. *Circulation, 48*: 185-194, 1973.

Ernsting, J., and Shephard, R.J.: Respiratory adaptations in congenital heart disease. *J Physiol, 112*: 332-343, 1951.

Estabrook, R.W., Franklin, M.R., and Hildebrandt, A.G.: Factors influencing the inhibitory effect of carbon monoxide on cytochrome P-450 catalyzed mixed function oxidation reactions. *Ann N Y Acad Sci, 174*: 218-232, 1970.

Ettre, L.S.: Application of gas chromatographic methods for air pollution studies. *J Air Pollut Control Assoc, 11*: 34-43, 1961.

Fabre, R., Truhaut, R., and Boudène, C.: Dosage de petites quantités de nicotine dans les atmosphères et les milieux biologiques animaux ou végétaux. Application a l'étude de l'absorption et de l'élimination de l'alcaloide chez les sujets exposés. *Ann Pharm Fr, 10*: 579-594, 1952.

Fabricant, N.D., and Rose, I.W.: Effect of cigarette smoking on the flicker fusion threshold of normal persons. *Eye, Ear, Nose and Throat Month, 30*: 541-543, 1951.

Faltings, K., Groth, W., and Harteck, P.: Photochemische Untersuchungen im Schumann-Ultraviolett Nr 7. Zur Photochemie der Kohlenoxyds. *Z Physik-Chem Abt B, 41*:15-22, 1938.

Farmer, C.J., and Crittenden, P.J.: A study of the carbon monoxide content of the blood of steel mill operatives. *J Ind Hyg, 11*: 329-335, 1929.

Fechter, L.D., and Annau, Z.: Effects of prenatal carbon monoxide exposure on neonatal rats. In Horvarth, M., and Frantek, E. (Eds.): *Adverse Effects of Environmental Chemicals and Psychotropic Drugs–Neurophysical and Behavioural Tests. Vol. 2.* New York, Elsevier, 1976, pp. 219-227.

Fechter, L.D., and Annau, Z.: Toxicity of mild prenatal carbon monoxide exposure.

Science, 197: 680-682, 1977.

Federal Aviation Administration Department of Transportation, and the National Institute for Occupational Safety and Health: *Health Aspects of Smoking in Transport Aircraft.* Washington, D.C., U.S. Govt. Print. Office, 1971.

Federal Highway Administration: *Carbon Monoxide: Toxic Gases in Heavy Duty Diesel Truck Cabs.* Washington, D.C., U.S. Govt. Print. Office, October, 1977.

Federal Minister of the Interior. *Technical Instructions for Maintaining Air Purity.* Bonn, August 28, 1974.

Federal Register, 38 (197): 28437, 1973.

Feldstein, M.: The colorimetric determination of blood and breath carbon monoxide. *J Forensic Sci, 10*: 35-42, 1965.

Feldstein, M.: Methods for the determination of carbon monoxide. In Stelman, Abraham (Ed.): *Progress in Chemical Toxicology. 3.* New York, Academic Press, 1967, pp. 99-119.

Fenn, W.O., and Cobb, D.M.: The burning of carbon monoxide by heart and skeletal muscle. *Am J Physiol, 102*: 393-401, 1932.

Ferris, B.G.: Health effects of exposure to low levels of regulated pollutants. A critical review. *J Air Pollut Control Assoc, 28*: 482-497, 1978.

Finkel, A.F.: Do air pollution effects outweigh benefits of highway jogging? *JAMA, 219*: 1224, 1972.

First, M.W., and Hinds, W.C. Ambient tobacco smoke measurements. *Am Ind Hyg Assoc J, 37*: 655-657, 1976.

Fischer, R.: The biological fabric of time. *Ann NY Acad Sci, 138*: 440-488, 1967.

Fischer, T., Weber, A., and Grandjean, E.: Air pollution due to tobacco smoke in restaurants. *Int Arch Occup Environ Health, 41:* 267-280, 1978.

Fisher, R.E.W.: Lung cancer propaganda. *Br Med J,*(i): 802, 1962.

Fitts, P.M., and Posner, M.I.: *Human Performance.* Belmont, California, Brooks/Cole Publishing Co., 1967.

Fleishman, E.A.: Dimensional analysis of psychomotor abilities. *J Exp Psychol, 48*: 437-454, 1954.

Flight Safety Foundation: *Flight Safety Facts and Analysis.* Arlington, Virginia, Flight Safety Foundation, Inc., 1973.

Fodor, G.G., and Winneke, G.: Effect of low CO concentrations on resistance to monotony and on psychomotor capacity. *Staub-Reinhalt Luft, 32*: 46-54, 1972.

Fold, G.E.: *Introduction to Environmental Physiology.* Philadelphia, Lea and Febiger, 1966.

Forbes, W.H.: CO uptake via the lungs. *Ann N Y Acad Sci, 174*: 72-75, 1970.

Forbes, W.H., Dill, D.B., De Silva, H., and Van Deventer, F.M.: The influence of moderate carbon monoxide poisoning upon the ability to drive automobiles. *J Ind Hyg Toxicol, 19*: 598-603, 1937.

Forbes, W.H., Sargent, F., and Roughton, F.J.W.: The rate of carbon monoxide uptake by normal men. *Am J Physiol, 143*: 594-608, 1945.

Forster, H.V., Dempsey, J.A., Birnbaum, M.L., et al.: Comparison of ventilatory responses to hypoxic and hypercapnic stimuli in altitude sojourning lowlanders, lowlanders residing at altitude and native altitude residents. *Fed Proc, 28*: 1274-1279, 1969.

Forster, R.E.: Reactions of carbon monoxide with heme proteins. In DuBois, A.B. (Ed.): *Effects of Chronic Exposure to Low Levels of Carbon Monoxide on Human Health, Behavior and Performance.* Washington, D.C., National Academy of Sciences and National Academy of Engineering, 1969, pp. 10-13.

Forster, R.E.: Carbon monoxide production by organisms. In DuBois, A.B. (Ed.): *Effects of Chronic Exposure to Low Levels of Carbon Monoxide on Human Health, Behavior, and Performance.* Washington, D.C., National Academy of Sciences and National Academy of Engineering, 1969b, pp. 15-16.

Forster, R.E.: Appendix: techniques for the analysis of carbon monoxide. In DuBois, A.B. (Ed.): *Effects of Chronic Exposure to Low Levels of Carbon Monoxide on Human Health, Behavior, and Performance.* Washington, D.C., National Academy of Sciences and National Academy of Engineering, 1969c, pp. 61-66.

Fowler, P.B.S.: Gangrene of the leg following carbon monoxide asphyxia. *Lancet, 1*: 240-241, 1954.

Frazer, W.M., and Stallybrass, C.O.: *Textbook of Public Health.* Edinburgh, E. and S. Livingstone, 1948.

Fristedt, B., and Akesson, B.: Health hazards from automobile exhausts at service facilities of multi-storey garages. *Hyg Rev, 60*: 112-118, 1971.

Furgang, F.A.: Carbon monoxide intoxication presenting as air embolism. *Aerospace Med, 43*: 785-786, 1972.

Gaensler, E.A., Cadigan, J.B., Ellicott, M.F., and Jones, R.H.: A new method for rapid precise determination of carbon monoxide in blood. *J Lab Clin Med, 49*: 945-957, 1957.

Gafafer, W.M.: *Occupational Diseases. A Guide to their Recognition.* Washington, D.C., U.S. Govt. Print. Office (DHEW Publication), 1964.

Gardner, R.S., Topping, D.L., and Mayes, P.A.: Immediate effects of carbon monoxide on the metabolism of chylomicron remnants by perfused rat liver. *Biochem Biophys Res Comm, 82:* 525-531, 1978.

Garner, L.L., Carl, E.F., and Grossman, E.E.: Effect of cigarette smoking on flicker fusion threshold. *Arch Ophthalmol, 51*: 642-655, 1954.

Garvey, D.J., and Longo, L.D.: Chronic low level maternal carbon monoxide exposure and fetal growth and development. *Biol Reprod, 19*: 8-14, 1978.

Gates, H.P., and Goldmuntz, L.A.: Expected decline in carboxyhaemoglobin levels as related to automibile carbon monoxide emission standards. *J Air Pollut Control Assoc, 26*: 891-892, 1976.

Geibel, J., Cannon, J., Campbell, D., and Traylor, T.G.: Model compounds for R state hemoglobins and T state hemoglobins. *J Am Chem Soc, 100:* 3575-3585, 1978.

General Electric: *Indoor-Outdoor Carbon Monoxide Pollution Study.* US EPA-RA-73-020, December, 1972.

Gennser, G., Marsal, K., and Brantmark, B.: Maternal smoking and fetal breathing movements. *Am J Obstet Gynecol, 123*: 861-867, 1975.

Georgii, H.W., Busch, E., and Weber, E.: *Investigation of the Temporal and Spatial Distribution of the Emission Concentration of Carbon Monoxide in Frankfurt/Main. Report of the Institute for Meteorology and Geophysics.* University of Frankfurt/Main, May, 1967.

Gettler, A.O., and Mattice, M.R.: The "normal" carbon monoxide content of the blood. JAMA, 100: 92-97, 1933.

Gibbs, G.W., Allard, D., Gray, R., and Corsillo, A.: Carbon monoxide and nitrogen dioxide exposure of Montreal longshoremen. *Can J Public Health, 64*: 147-153, 1973.

Gibson, O.H.: The kinetics of reaction between haemoglobin and gases. *Prog Biophys 9*: 1-53, 1959.

Gibson, O.H., and Ainsworth, S.: Photosensitivity of haem compounds. *Nature, 180:* 1416-1417, 1957.

Gibson, O.H., and Greenwood, C.: The spectra and some properties of cytochrome oxidase components. *J Biol Chem, 239*: 586-590, 1964.

Gilli, G., Corrao, G., and Scursaton, E.: Valori Relativi alla Carbossiemoglobina in un Gruppo di Popolazione. Analisi di Alcune Variabili. *Giorn di Ricerca e Diag di Lab, 6*: 389-396, 1979.

Gilmore, T.M., and Hanna, T.R.: The need for representative ambient air carbon monoxide sampling. *J Air Pollut Control Assoc, 26*: 965-967, 1976.

Ginsberg, M.D., and Myers, R.E.: Fetal brain damage following maternal carbon monoxide intoxication. An experimental study. *Acta Obstet Gynecol Scand, 53*: 309-317, 1974.

Ginsberg, M.D., and Myers, R.E.: Fetal brain injury after maternal carbon monoxide intoxication. Clinical and neuropathologic aspects. *Neurology, 26*: 15-23, 1976.

Glaser, E.M.: *The Physiological Basis of Habituation.* London, Oxford University Press, 1966.

Glass, R.I.: A perspective on environmental health in the U.S.S.R. *Arch Environ Health, 30*: 391-395, 1975.

Gliner, J.A., Raven, P.B., Horvath, S.M., et al.: Man's physiologic response to long-term work during thermal and pollutant stress. *J Appl Physiol, 39*: 628-632, 1975.

Goddard, R.: *International Symposium on the Effects of Altitude on Physical Performance.* Chicago, Athletic Institute, 1967.

Godin, G., and Shephard, R.J.: On the course of carbon monoxide uptake and release. *Respiration, 29:* 317-329, 1972.

Godin, G., Wright, G.R., and Shephard, R.J.: Urban exposure to carbon monoxide. *Arch Environ Health, 25*: 305-313, 1972.

Goldbaum, L.R.: Is carboxyhemoglobin concentration the indicator of carbon monoxide toxicity? *Leg Med Ann, 77*: 165-170, 1977.

Goldberg, H.D., and Chappell, M.N.: Behavioral measure of effect of carbon monoxide on rats. *Arch Environ Health, 14:* 671-677, 1967.

Goldberg, L.: Evaluation of benefit vs. potential hazard in environmental toxicology. In *Pharmacology and the Future of Man. Vol 2. Proceedings of the 5th International Congress of Pharmacology.* Basel, Karger, 1972, pp. 13-23.

Goldman, A., Murcray, D.G., Murcray, F.H., et al.: Vertical distribution of CO in the atmosphere. *J Geophys Res, 78*: 5273-5283, 1973.

Goldsmith, J.R.: Some implications of ambient air quality standards. *Arch Environ Health, 4*: 151-167, 1962.

Goldsmith, J.R.: In Discussion: Epidemiologic studies of chronic respiratory diseases. *Arch Environ Health, 10*: 383-388, 1965.

Goldsmith, J.R.: Contribution of motor vehicle exhaust, industry and cigarette smoking to community carbon monoxide exposures. *Ann NY Acad Sci, 174*: 122-134, 1970.

Goldsmith, J.R., Terzaghi, J., and Hackney, J.D.: Evaluation of fluctuating carbon monoxide exposures. *Arch Environ Health, 7*: 647-663, 1963.

Goldsmith, J.R., and Landaw, S.A.: Carbon monoxide and human health. *Science, 162*: 1352-1359, 1968.

Goldsmith, J.R., Beard, R.R., and Dinman, B.D.: Epidemiological appraisal of carbon monoxide effects. In DuBois, A.B. (Ed.): *Effects of Chronic Exposure to Low Levels of Carbon Monoxide on Human Health, Behavior and Performance.* Washington, D.C., National Academy of Sciences and National Academy of Engineering, 1969, pp. 40-60.

Goldsmith, J.R., and Aronow, W.S.: Carbon monoxide and coronary heart disease: a review. *Environ Res, 10*: 236-248, 1975.

Goldstein, B.D., Goldring, R.M., and Amorosi, E.L.: *Investigation of Mechanisms of Oxygen Delivery to the Tissues in Individuals with Chronic Low-Grade Carbon Monoxide Exposure.* New York University School of Medicine, Report to Coordinating Research Council, Contract CAPM-8-68, 1975, pp. 1-20.

Goldstein, E.: Re-evaluation of the United States Air Quality standard for nitrogen dioxide. *Rev Environ Health, 2*: 5-37, 1975.

Goldstein, H.: Factors influencing the height of seven year old children—results from the National Child Development Study. *Hum Biol, 43*: 92-111, 1971.

Gorbatow, O., and Noro, L.: On acclimatization in connection with acute carbon monoxide poisonings. *Acta Physiol Scand, 15*: 77-87, 1948.

Gordon, G.S., and Rogers, R.L.: *A Report of Medical Findings of Project Monoxide. Redmond Memorial Fund, International Association of Firefighters.* Washington, Merkel Press, 1969.

Gorodinsky, S.M., Sedov, A.V., Mazin, A.N., et al.: Effect of reduced barometric pressure on the elimination of gaseous and volatile metabolites of suited man. *Cosm Biol Med, 2*: 78-82, 1970.

Göthe, C.J., Frisfedt, B., Sundell, L., et al.: Carbon monoxide hazard in city traffic. *Arch Environ Health, 19*: 310-314, 1969.

Göthert, M., Lutz, F., and Malorny, G. Carbon monoxide partial pressure in tissue of different animals. *Environ Res, 3*: 303-309, 1970.

Graham, J.I.: The normal production of carbon monoxide in coal-mines. *Trans Instn Min Engrs, London, 60*: 222, 1920.

Griggs, T.R.: The role of exertion as a determinant of carboxyhaemoglobin accumulation in firefighters. *J Occup Med, 19*: 759-761, 1977.

Grimmer, G., Boehnke, H., and Harke, H.P.: Zum Problem des Passivrauchens: Aufnahme von polycyclischen aromatischen Kohlenwasserstoffen durch Einatmen von zigarettenrauchhaltiger Luft. *Int Arch Occup Environ Health, 40*: 93-99, 1977.

Groll-Knapp, E., Wagner, H., Hanck, H., and Harder, M.: Effects of low carbon monoxide concentrations on vigilance and computer-analyzed brain potentials. *Staub-Reinhalt Luft, 32*: 64-68, 1972.

Grut, A.: *Chronic Carbon Monoxide Poisoning*. Copenhagen, Ejnar Munksgaard, 1949.

Grut, A., Åstrup, P., Challen, P.J.R., and Gerhardsson, G.: Threshold limit values for carbon monoxide. *Arch Environ Health, 21*: 542-544, 1970.

Guest, A.D., Duncan, C., and Lawther, P.J.: Carbon monoxide and phenobarbitone: a comparison of effects on auditory flutter fusion threshold and critical flicker fusion threshold. *Ergonomics, 13*: 587-594, 1970.

Guidotti, T.L.: The higher oxides of nitrogen: inhalation toxicology. *Environ Res, 15*: 443-472, 1978.

Guillerm, R., Badré, R., Hee, J., and Masurel, G.: Composition de la fumée de tabac. Analyse des facteurs de nuisance. *Revue de Tuberculose et de Pneumologie, 36*: 187-208, 1972.

Guillerm, R., Radziszewski, E., and Caille, J.E.: Effects of carbon monoxide on performance in a vigilance task (automobile driving). In Thornton, R.E. (Ed.): *Smoking Behavior. Physiological and Psychological Influences.* Edinburgh, Churchill-Livingstone, 1978, pp. 148-155.

Gunther, H.: Zur Pathogenese der Kohlenoxydvergiftung. *Z Klin Med, 92*: 41-53, 1921.

Haag, W.M.: Methods to reduce carbon monoxide levels at the workplace. *Prev Med, 8*: 369-378, 1979.

Haagen-Smit, A.J.: Carbon monoxide levels in city driving. *Arch Environ Health, 12*: 548-551, 1966.

Hackney, J.D., Kaufman, G.A., Lashier, H., and Lynn, K.: Rebreathing estimate of carbon monoxide hemoglobin. *Arch Environ Health, 5*: 300-307, 1962.

Hackney, J.D., Linn, W.S., Mohler, J.G., et al.: Experimental studies on human health. Effects of air pollutants. II. Four-hour exposure to ozone alone and in combination with other pollutant gases. *Arch Environ Health, 30:* 379-384, 1975.

Häggendal, E., Nilsson, N.J., and Norbach, B.: Effect of blood corpuscle concentration on cerebral blood flow. *Acta Chir Scand Suppl, 364*: 3-12, 1966.

Haider, M., Groll-Knapp, E., Holler, H., et al.: Electro-physiological and behaviour data and clinical relevance. In Funkel, A.J., and Duel, W.C. (Ed.): *Clinical Implications of Air Pollution Research.* Acton, Massachusetts, Publishing Sciences Group, Inc., 1975, pp. 217-232.

Hahn-Pedersen, A.: *Investigations of Prolonged Heavy Carbon Monoxide—Hypoxia in Albino-Female Rats.* Doctoral dissertation, University of Copenhagen, 1962.

Haldane, J.S.: The relation of the action of carbonic oxide to oxygen tension. *J Physiol, 18:* 201-217, 1895a.

Haldane, J.S.: The action of carbonic oxide on man. *J Physiol, 18:* 430-462, 1895b.

Haldane, J.S.: A method of detecting and estimating carbonic oxide in air. *J Physiol, 18:* 463-469, 1895c.

Haldane, J.S.: Investigations on the nature and sources of the suffocative gas met with in coal mines: together with further observations on the black-damp of coal-mines. *Trans Inst Min Engrs, London, 11:* 265, 1895d.

Haldane, J.S.: *Report on the Causes of Death in Colliery Explosions and Fires.* U.K. Parliamentary Paper *C,* 8112, 1896a.

Haldane, J.S.: Poisoning by gas in sewers. *Lancet,* (i): 220-224, 1896a.

Haldane, J.S.: Report on the composition of the air in the Metropolitan and other railway tunnels and on means of artificial ventilation of tunnels. Cited by J.S. Haldane and J.G. Priestley in *Respiration.* Oxford, Clarendon Press, 1897.

Haldane, J.S.: *Report of the Water Gas Committee.* British Parliamentary Paper, 1899.

Haldane, J.S.: Self-contained rescue apparatus and smoke helmets: second report of the Doncaster Coal-Owner's (Gob-fire research) Committee. *Trans Inst Min Engrs, London, 48:* 550-585, 1915.

Haldane, J.S.: Acclimatisation to high altitudes. *Physiol Rev, 7:* 363-384, 1927.

Haldane, J., and Smith, J.L.: The oxygen tension of arterial blood. *J Physiol, 20:* 497-517, 1896.

Haldane, J., and Smith, J.L.: The absorption of oxygen by the lungs. *J Physiol, 22:* 231-258, 1898.

Haldane, J.S., and Meachem, F.G.: Observations on the relation of underground temperature and spontaneous fires in the coal to oxidation and to the causes which favour it. *Trans Inst Min Engrs, London, 16:* 457, 1898.

Haldane, J.S., and Makgill, R.H.: The formation of black-damp and of carbon monoxide from coal at ordinary temperatures. *Trans Inst Min Engrs, London, 85:* 172, 1933.

Haldane, J.S., and Priestley, J.G.: *Respiration,* 2nd ed. Oxford, Clarendon Press, 1935.

Hall, D.A., Miller, F.A., Riley, E.C., and Scherberger, R.F.: Evaluation of the carbon monoxide hazard from indoor use of propane-fueled fork lift trucks. *Am Ind Hyg Assoc Q, 18:* 355-359, 1957.

Hall, S.A.: Airborne contaminants. In Schilling, R.S.F. (Ed.): *Occupational Health Practice.* London, Butterworths, 1973, pp. 288-308.

Halperin, M.H., McFarland, R.A., Niven, J.I., and Roughton, F.J.W.: The time course of the effects of carbon monoxide on visual thresholds. *J Physiol, 146:* 583-593, 1959.

Hamming, W.J., MacPhee, R.D., and Taylor, J.R.: Contaminant concentrations in the atmosphere of Los Angeles County. *J Air Pollut Control Assoc, 10:* 7-16, 1960.

Hampson, R.F., and Garvin, D.: *Chemical Kinetic and Photochemical Data for Modeling Atmospheric Chemistry.* Washington, D.C., U.S. Govt. Print. Office (NBS Techn. Note 886), 1975, pp. 1-112.

Hanks, T.G.: Human performance of a psychomotor test as a function of exposure to carbon monoxide. *Ann NY Acad Sci, 174:* 421-424, 1970.

Hanson, H.B., and Hastings, A.B.: The effect of smoking on the carbon monoxide content of blood. *JAMA, 100:* p. 1481, 1933.

Hanst, P.L., Lefohn, A.S., and Gay, B.W.: Detection of atmospheric pollution at parts per billion levels by infrared spectroscopy. *J Appl Spectr, 27:* 188-198, 1973.

Harke, H.P.: Zum Problem des "Passiv-Rauchens." (The problem of "passive smoking"). *Munch Med Wschr, 112:* 2328-2334, 1970.

Harke, H.P., and Bleichert, A.: Zum Problem des "Passiv-rauchens." *Int Arch Arbeitsmed, 29:* 312-322, 1972.

Harke, H.P., Baars, A., Frahm, B., et al.: Relation between smoke constituents and number of smoked cigarettes, time and room volume. *Int Arch Arbeitsmed, 29:* 323-329, 1972.

Harke, H.P., Liedl, W., and Denkers, D.: The problem of passive smoking II. Investigations of CO levels in the automobile after cigarette smoking. *Int Arch Arbeitsmed, 33:* 207-220, 1974.

Harke, H.P., and Peters, H.: The problem of passive smoking III. The influence of smoking on the CO concentrations in driving automobiles. *Int Arch Arbeitsmed, 33:* 221-229, 1974.

Harke, H.P., Schmeltz, I., Hoffman, D., and Wynder, E.L.: The influence of tobacco smoke on indoor atmospheres. *Prev Med, 4:* 373-377, 1975.

Harke, H.P., and Peters, H.: The importance of smoking for the CO concentration in cars in traffic. *Arch Hygiene und Bakteriol, 141:* 383-400, 1977.

Harper, P.V.: A new spectrophotometric method for the determination of CO in the blood. *J Lab Clin Med, 40:* 634-640, 1952.

Hart, J.S.: Climatic and temperature induced changes in the energetics of homeotherms. *Rev Can Biol, 16:* 133-141, 1957.

Hartmann, H.F.: Carbon monoxide—occurrence, concentration, effects on man—a review. *Clean Air, 6:* 2-7, 1972.

Hartridge, H.: A spectroscopic method of estimating carbon monoxide. *J Physiol, 44:* 1-21, 1912a.

Hartridge, H.: The action of various conditions on carbon monoxide haemoglobin. *J Physiol, 44:* 22-23, 1912b.

Hartridge, H.: CO in tobacco smoke. *J Physiol, 53:* 83-84, 1920.

Hass, G.C., and Brubacher, M.L.: A test procedure for motor vehicle exhaust emissions. *J Air Pollut Control Assoc, 12:* 505-509 and 543, 1962.

Hatch, T.F.: Changing objectives in occupational health. *Am Ind Hyg Assoc J, 23:* 1-7, 1962.

Hatch, T.F.: Permissible levels of exposure to hazardous agents in industry. *J Occup Med, 14:* 134-137, 1972.

Hatch, T.F.: Criteria for hazardous exposure limits. *Arch Environ Health, 27:* 231-235, 1973.

Hedinger, C.: Zur Pathologie der Skelettmuskulatur. Muskelveränderungen bei Kohlenmonoxydvergiftung. Ihre Beziehungen zum Verschüttungssyndrom. *Schweiz Med Wschr, 78:* 145-151, 1948.

Heliovaara, M., Karvonen, M.J., Vilhunen, R., and Punsar, S.: Smoking, carbon monoxide and atherosclerotic diseases. *Br Med J,* (i): 268-270, 1978.

Hellman, L.M., and Pritchard, J.A.: *Williams Obstetrics,* 14th ed. New York, Appleton-Century-Crofts, 1971.

Henderson, M., and Apthorp, G.H.: Rapid method for estimation of carbon monoxide in blood. *Br Med J,* (ii): 1853-1854, 1960.

Henderson, Y., and Haggard, H.W.: The elimination of carbon monoxide from the blood after a dangerous degree of asphyxiation, and a therapy for accelerating elimination. *J Pharmacol Exp Ther, 16:* 11-20, 1921.

Henderson, Y., Haggard, H.W., Teague, M.C., et al.: Physiological effects of automobile exhaust gases and standards of exposure for brief periods. *J Ind Hyg, 3:* 79-92, 137-146, 1921.

Henderson, Y., and Haggard, H.W.: Health hazard from automobile exhaust gas in city streets, garages and repair shops. *JAMA, 81:* 385-391, 1923.

Hernberg, S., Karava, R., Koskela, R.S., and Luoma, K.: Angina pectoris, ECG findings and blood pressure of foundry workers in relation to carbon monoxide exposure. *Scand J Work Environ Health, 2* (Suppl. 1): 54-63, 1976.

Heron, H.J.: The effects of smoking during pregnancy. A review with a preview. *NZ Med J, 61:* 545-548, 1962.

Hersch, P.: Galvanic analysis. In Reilley, C.A. (Ed.): *Advances in Analytical Chemistry and Instrumentation. Vol. 3.* New York, Interscience Publishers, 1964, pp. 183-249.

Hess, H.A.: Zum Problem des Passivrauchens. Bemerkung zu H.P. Harke. *Munch Med Wschr, 113:* 705-706, 1971.

Heuss, J.M., Nebel, G.J., and Colucci, J.M.: National air quality standards for automobile pollutants. A critical review. *J Air Pollut Control Assoc, 21:* 535-544, 1971.

Hexter, A.C., and Goldsmith, J.R.: Carbon monoxide: Association of community air pollution with mortality. *Science, 172:* 265-267, 1971.

Hickey, R.J., Clelland, R.C., and Boyce, D.E.: Carbon monoxide: smoking, air pollution, cardiovascular disease and physiological homeostasis. (Letter). *Lancet,* (ii): 571-572, 1973.

Hickey, R.J., Clelland, R.C., Boyce, D.E., and Harner, E.B.: Carbon monoxide and smoking. (Letter). *Lancet,* (i): 409-410, 1974.

Hill, D.W., and Powell, T.: *Non-Dispersive Infra-Red Gas Analysis in Science, Medicine and Industry.* London, Adam Hilger Ltd., 1968.

Hill, E.P., Hill, J.R., Power, G.G., and Longo, L.D.: Carbon monoxide exchanges between the human fetus and mother: a mathematical model. *Am J Physiol, 232:* H311-H323, 1977.

Hirtzel, C.S., and Quon, J.E.: Statistical dependence of hourly carbon monoxide measurements. *J Air Pollut Control Assoc, 29:* 161-163, 1979.

Hlastala, M.P., McKenna, H.P., Franada, R.L., and Detter, J.C.: Influence of carbon monoxide on haemoglobin-oxygen binding. *J Appl Physiol, 41:* 893-899, 1976.

Hody, G.L., and Schane, W.P.: *Preliminary Carbon Monoxide Measurements in Armed Helicopters.* Fort Rucker, Alabama, U.S. Army Aeromedical Research Lab (Report 66-1), 1966.

Hoffmann, D., and Wynder, E.L.: Smoke of cigarettes and little cigars: an analytic comparison. *Science, 178:* 1197-1199, 1972.

Hoffman, D.J., and Campbell, K.I.: Embryo toxicity of irradiated and non-irradiated automotive exhaust and carbon monoxide. *Environ Res, 15:* 100-107, 1978.

Hofreuter, D.H., Calcott, E.J., and Xintaras, C.: Carboxyhemoglobin in men exposed to carbon monoxide. *Arch Environ Health, 4:* 81-85, 1962.

Holland, R.A.B.: Rate of O_2 dissociation from O_2Hb and relative combination rate of CO and O_2 in mammals at 37° C. *Respir Physiol, 7:* 30-42, 1969a.

Holland, R.A.B.: Rate at which CO replaces O_2 from O_2Hb in red cells of different species. *Respir Physiol, 7:* 43-63, 1969b.

Horvath, S.M.: In *Proceedings of Conference on Health Effects of Air Pollutants. Assembly of Life Sciences NAS-NRC.* Washington, D.C., U.S. Govt. Print. Office (Serial#93-15), 1973, pp. 127-139.

Horvath, S.M., and Roughton, F.J.W.: Improvements in the gasometric estimation of carbon monoxide in the blood. *J Biol Chem, 144:* 747-755, 1942.

Horvath, S.M., Dahms, T.E., and O'Hanlon, J.F.: Carbon monoxide and human vigilance—a deleterious effect of present urban concentrations. *Arch Environ Health, 23:* 343-347, 1971.

Horvath, S.M., Raven, P.B., Dahms, T.E., and Gray, D.J.: Maximal aerobic capacity at different levels of carboxyhemoglobin. *J Appl Physiol, 38:* 300-303, 1975.

Hosein, H.R., and Bouhuys, A.: Possible environmental hazards of gas cooking. *Br Med J,* (i): 125, 1979.

Hosey, A.D.: Priorities in developing criteria for "breathing air" standards. *J Occup Med, 12:* 43-46, 1970.

Hosko, M.J.: The effect of carbon monoxide on the visual evoked response of man. *Arch Environ Health, 21:* 174-180, 1970.

Howlett, L., and Shephard, R.J.: Carbon monoxide as a hazard in aviation. *J Occup Med, 15:* 874-877, 1973.

Hughes, E.E.: Development of standard reference materials for air quality measurement. *ISA Trans, 14:* 281-291, 1975.

Hugod, C., Hawkins, L.H., Kjeldsen, K., et al.: Effect of carbon monoxide exposure on aortic and coronary intimal morphology in the rabbit. A re-evaluation. *Atherosclerosis, 30:* 333-342, 1978.

Hugod, C., Hawkins, L.H., and Åstrup, P.: Exposure of passive smokers to tobacco smoke constituents. *Int Arch Occup Environ Health, 42:* 21-29, 1978.

Hurn, R.W.: Comprehensive analyses of automotive exhausts. *Arch Environ Health, 5:* 592-596, 1962.

Hurtado, A.: Acclimatization to high altitudes. In Weihe, W.H. (Ed.): *The Physiological Effects of High Altitude.* New York, MacMillan, 1964.

Hurtado, A.: The influence of high altitude on physiology. In Porter, R., and Knight, J. (Eds.): *High Altitude Physiology: Cardiac and Respiratory Aspects.* Edinburgh Churchill-Livingstone, 1971.

Ingenito, A.J.: Effects of cigarette smoke inhalation on the cat electroretinogram: comparisons with nicotine and carbon monoxide. *J Pharmacol Exp Ther, 211:* 647-655, 1979.

Ingenito, A.J., and Durlacher, L.: Effects of carbon monoxide on the b-wave of the cat electroretinogram: comparisons with nitrogen hypoxia, epinephrine, vasodilator drugs and changes in respiratory tidal volume. *J Pharmacol Exp Ther, 211:* 638-646, 1979.

Inman, R.E., Ingersoll, R.B., and Levy, E.A.: Soil: A natural sink for carbon monoxide. *Science, 172:* 1229-1231, 1971.

Inoue, R., and Watanabe, Y.: *Statistical Properties of Sulfur Dioxide Concentrations in Continuous Monitoring Stations in Japan.* Abstracts, 4th International Clean Air Congress, Tokyo, 1977.

International Labour Office. *Permissible Levels of Toxic Substances in Air in Working Environment.* 6th Session, Joint ILO/WHO Committee on Occupational Health. Geneva, Occupational Health and Safety Series (*20*), 1970.

Jacobs, M.B.: *The Analytical Chemistry of Industrial Poisons, Hazards and Solvents,* 2nd ed. New York, Interscience, 1949.

Jacobs, M.B.: *The Chemical Analysis of Air Pollutants.* New York, Interscience, 1960.

Jacobs, M.B., Braverman, M.M., and Hockheiser, S.: Continuous determination of carbon monoxide and hydrocarbons in air by a modified infra-red analyzer. *J Air Pollut Control Assoc, 9:* 110-114, 1959.

Jaffe, L.S.: Ambient carbon monoxide and its fate in the atmosphere. *J Air Pollut Control Assoc, 18:* 534-540, 1968.

Jaffe, L.S.: Sources, characteristics and fate of atmospheric carbon monoxide. *Ann NY Acad Sci, 174:* 76-88, 1970.

Jaffe, L.S.: Carbon monoxide in the biosphere: sources, distribution and concentrations. *J Geophys Res, 87:* 5293-5305, 1973.

James, G.: Health challenges today. *Am Rev Respir Dis, 90:* 349-358, 1964.

Jarvis, M.J., Russell, A.H., and Saloojec, Y.: Expired air carbon monoxide: a simple breath test of tobacco smoke intake. *Br Med J,* (i): 484-485, 1980.

Jermini, C., Weber, A., and Grandjean, E.: Quantitativ Bestimmung verschiedener Gasphasen -komponenten des Nebenstromrauches von Zigaretten in der Raumluft als Beitrag zum Problem des Passiv-rauchens. *Int Arch Occup Environ Health, 36:* 169-181.

Joels, N., and Pugh, L.G.C.E.: The carbon monoxide dissociation curve of human blood. *J Physiol, 142:* 63-77, 1958.

Johnson, B.L., Cohen, H.H., Struble, R., et al.: Field evaluation of carbon monoxide exposed toll collectors. In Xintaras, C., Johnson, B.L., and deGroot, I. (Eds.): *Behavioral Toxicology.* Washington, D.C., U.S. Govt. Print. Office (DHEW Publication NIOSH Report 74-126), 1974, pp. 306-328.

Johnson, K.L., Dworetzky, L.H., and Heller, A.N.: Carbon monoxide and air pollution from automobile emissions in New York City. *Science, 160:* 67-68, 1968.

Jones, J.G.: The use of respirators and breathing apparatus in the iron and steel industry. In Davies, C.N. (Ed.): *Design and Use of Respirators.* Oxford, Pergamon Press, 1961, pp. 127-135.

Jones, J.G., and Walters, D.H.: A study of carboxyhaemoglobin levels in employees at an integrated steelworks. *Ann Occup Hyg, 5:* 221-230, 1962.

Jones, J.G., and Sinclair, A.: Arterial disease among blast furnace workers. *Ann Occup Hyg, 18:* 15-20, 1975.

Jones, R.A., Strickland, J.A., Stunkard, J.A., and Siegel, J.: Effects on experimental animals of long-term inhalation exposure to carbon monoxide. *Toxicol Appl Pharmacol, 19:* 46-53, 1971.

Jones, R.D., Commins, B.T., and Cernik, A.A.: Blood lead and carboxyhaemoglobin levels in London taxi drivers. *Lancet,* (ii): 302-303, 1972.

Jones, R.H., Ellicott, M.F., Cadigan, J.B., and Gaensler, E.A.: The relationship between alveolar and blood carbon monoxide concentrations during breath holding. *J Lab Clin Med, 51:* 553-564, 1958.

Junge, C.E.: *Air Chemistry and Radioactivity.* New York, Academic Press, 1963.

Kaempe, B., and Dalgaard, J.B.: Drugs, alcohol and carbon monoxide in traffic accidents. *Ugeskr Laeger, 141:* 1036-1040, 1979.

Kagan, A., Harris, B.R., Winkelstein, W., et al.: Epidemiologic studies of coronary heart disease and stroke in Japanese men living in Japan, Hawaii and California: Demographic, physical, dietary and biochemical characteristics. *J Chronic Dis, 27:* 345-364, 1974.

Kahn, A., Rutledge, R.B., Davis, G.L., et al.: Carboxyhaemoglobin sources in the metropolitan St. Louis population. *Arch Environ Health, 29:* 127-135, 1974.

Kalmaz, E.V., Cauter, L.W., Nau, C.A., and Hampton, J.W.: Effect of carbon monoxide exposure on blood fibrinolytic activity in rabbits. *Environ Res, 14:* 194-202, 1977.

Kandall, S.R., Landaw, S.A., and Thaler, M.M.: Carboxyhemoglobin exchange between donors and recipients of blood transfusions. *J Pediatr, 52:* 716-718, 1973.

Kannel, W.B., and Castelli, W.P.: Significance of nicotine, carbon monoxide and other smoke components in the development of cardiovascular disease. In Wynder, E.L., Hoffmann, D., and Gori, G.B. (Eds.): *Modifying the Risk for the Smoker. Vol. I. Proceedings of the 3rd World Conference on Smoking and Health.* Washington, D.C., U.S. Govt. Print. Office (DHEW Publication NIH 76-1221), 1976, pp. 369-381.

Karnik, A.S., and Coin, E.J.: Carbon monoxide kinetics following simulated cigarette smoking. *Arch Environ Health, 35:* 146-147, 1980.

Katz, M., and Drowley, W.B.: Canadian activities in ambient air quality criteria and development of standards. *J Air Pollut Control Assoc, 16:* 131-134, 1966.

Katz, S.H., and Frevert, H.W.: Carbon monoxide in two large garages. *J Ind Eng Chem, 20:* 31-36, 1928.

Kelley, J.S., and Sophocleus, G.J.: Retinal hemorrhages in subacute carbon monoxide poisoning. Exposures in homes with blocked furnace flues. *JAMA, 239:* 1515-1517, 1978.

Kellogg, R.H.: The role of CO_2 in altitude acclimatization. In Cunningham, D.J.C., and Lloyd, B.B. (Eds.): *The Regulation of Human Respiration.* Oxford, Blackwell Scientific, 1963, pp. 379-395.

Kennedy, A.C., and Valtis, D.J.: The oxygen dissociation curve in anemia of various types. *J Clin Invest, 33:* 1372-1381, 1954.

Kernohan, J.C.: Kinetics of the reactions of two sheep haemoglobins with oxygen and carbon monoxide. *J Physiol, 155:* 580-588, 1961.

Kerr, D.J.A.: *Forensic Medicine,* 4th ed. London, Adam and Charles Black, 1946.

Kersten, H.: Kohlenmonoxidhaltige Atemluft Fordert Arterienverkalkung. *Krankenpflege, 6:* 161, 1976.

Kessler, M., Lang, H., Sinagowitz, et al.: Homeostasis of oxygen supply in liver and kidney. In Bicher, H.I., and Bruley, D.F. (Eds.): *Oxygen Transport to Tissue: Instrumentation, Methods and Physiology.* New York, Plenum Press, 1973, pp. 351-360.

Killick, E.M.: The acclimatization of the human subject to atmospheres containing low concentrations of carbon monoxide. *J Physiol, 87:* 41-55, 1936.

Killick, E.M.: The acclimatization of mice to atmospheres containing low concentrations of carbon monoxide. *J Physiol, 91:* 279-292, 1937.

Killick, E.M.: Carbon monoxide anoxemia. *Physiol Rev, 20:* 313-344, 1940.

Killick, E.M.: The nature of the acclimatization occurring during repeated exposures of the human subject to atmospheres containing low concentrations of carbon monoxide. *J Physiol, 107:* 27-44, 1948.

Killick, E.M., and Marchant, J.V.: Resuscitation of dogs from severe acute carbon monoxide poisoning. *J Physiol, 147:* 274-298, 1959.

Kindwall, E.P.: Carbon monoxide and cyanide poisoning. In Davis, J.C., and Hunt, T.K. (Eds.): *Hyperbaric Oxygen Therapy.* Bethesda, Maryland, Undersea Medical Society, 1977, Chapter 13, pp. 177-190.

Kjeldsen, K., and Damgaard, F.: Influence of prolonged carbon monoxide exposure and altitude hypoxia on serum lipids in man. *Scand J Clin Lab Invest, 22, Suppl. 103:* 16-19, 1968a.

Kjeldsen, K., and Damgaard, F.: Influence of prolonged carbon monoxide exposure and high altitude on the composition of blood and urine in man. *Scand J Clin Lab Invest, 22, Suppl. 103:* 20-25, 1968b.

Kjeldsen, K., Wanstrup, J., and Åstrup, P.: Enhancing influence of arterial hypoxia on the development of atheromatosis in cholesterol-fed rabbits. *J Atheroscler Res, 8:* 835-845, 1968.

Kjeldsen, K., Åstrup, P., and Wanstrup, J.: Ultra-structural intimal changes in the rabbit aorta after a moderate carbon monoxide exposure. *Atherosclerosis, 16:* 67-82, 1972.

Kjeldsen, K., Thomsen, H.K., and Åstrup, P.: The effects of carbon monoxide on myocardium: ultrastructural changes in rabbits after a moderate, chronic exposure. *Circ Res, 34*: 339-348, 1974.

Klausen, K., Rasmussen, B., Gjellerod, H., et al.: Circulation, metabolism and ventilation during prolonged exposure to CO and to high altitude. *Scand J Clin Lab Invest, 22, Suppl. 103:* 26-38, 1968.

Kleiner, B., and Spengler, J.: Carbon monoxide exposure of Boston bicyclists. *J Air Pollut Control Assoc, 26*: 147-149, 1976.

Klemisch, H.J., Chevalier, H.J., Harke, J.P., and Dontevill, W.: Uptake of carbon monoxide in blood of miniature pigs and other mammals. *Toxicology, 3*: 301-310, 1975.

Klendshøj, N.C., Feldstein, M., and Sprague, A.L.: The spectrophotometric determination of carbon monoxide. *J Biol Chem, 183*: 297-303, 1950.

Klocke, R.A.: Pulmonary excretion of absorbed gases. In Lee, D.H.K. (Ed.): *Handbook of Physiology. Section 9. Reactions to Environmental Agents.* Washington, D.C., American Physiological Society, 1977, pp. 555-562.

Komatsu, F.: Shinshu myocardosis. *Digest of Science of Labour, 10:* 315-318, 1955.

Kreuzer, F.: Transport of O_2 and CO_2 at altitude. In Margaria, R. (Ed.): Exercise at altitude. Dordrecht, Excerpta Medica Fdn., 1967, pp. 149-158.

Krotov, J.A.: Maximum permissible concentrations of harmful substances in air and water. Cited by WHO, 1975.

Kubic, C.V.L., and Anders, M.W.: Metabolism of dihalomethanes to carbon monoxide. Part I. In vivo studies. *Drug Metab, 2*: 53-57, 1974.

Kuller, L.H., Radford, E.P., Swift, D., et al.: Carbon monoxide and heart attacks. *Arch Environ Health, 30*: 477-482, 1975.

Kuller, L., Perper, J., and Cooper, M.: Sudden and unexplained death due to arterio sclerotic heart disease. In Oliver, M.F. (Ed.): *Modern Trends in Cardiology, Vol 3.* London, Butterworths, 1975, pp. 292-332.

Kurt, T.L., Mogielnicki, R.P., and Chandler, J.E.: Association of the frequency of acute cardiorespiratory complaints with ambient levels of carbon monoxide. *Chest, 74*: 10-14, 1978.

Kyryacos, G., and Boord, C.E.: Separation of hydrogen, oxygen, nitrogen, methane and carbon monoxide by gas absorption chromatography. *Anal Chem, 29:* 787-788, 1957.

Laham, S., Potvin, M. and Schrader, K.: Toxicological studies on dichloromethane, a solvent stimulating carbon monoxide poisoning. *Toxicol Eur Res, 1:* 63-73, 1978.

Lahiri, S., and Milledge, J.S.: Relative respiratory insensitivity of Himalayan Sherpa altitude residents to hypoxia at 4880 m and at sea level. In Torrance, R.W. (Ed.): *Arterial Chemoreceptors.* Oxford, Blackwell Scientific, 1968, pp. 387-392.

Lahiri, S., Brody, J.S., Motoyama, E.K., and Velasquez, T.M.: Regulation of breathing in newborns at high altitude. *J Appl Physiol, 44:* 673-678, 1978.

Lambert, J.L., Tschorn, R.R., and Hamlin, P.A.: Determination of carbon monoxide in blood. *Anal Chem, 44:* 1529-1530, 1972.

Landis, C.: Determinants of the critical flicker-fusion threshold. *Physiol Rev, 34*: 259-286, 1954.

Lanino, P.G.: Safety and health in foundries. *Fonderia, 13:* 561-566 (Milano), 1964.

Larsen, R.I., and Konopinski, V.J.: Summer tunnel air quality. *Arch Environ Health, 5:* 597-608, 1962.

Larsen, R. I., and Burke, H.: *Ambient Carbon Monoxide Exposure.* Paper 69-167 presented at 62nd Annual Meeting of Air Pollution Control Association. Cited by Coburn, R.F. (1977).

Larson, G.P., Chipman, J.C., and Kauper, E.K.: Study of the distribution and effects of auto exhaust gases. *J Air Pollut Control Assoc, 5*: 84-90, 1955.

Larson, P.S., Finnegan, J.K., and Haag, H.B.: Observations on the effect of cigarette smoking on the fusion frequency of flicker. *J Clin Invest, 29:* 483-485, 1950.

Laties, V.G., and Merigan, W.H.: Behavioural effects of carbon monoxide on animals and man. *Ann Rev Pharmacol Toxicol, 19:* 357-392, 1979.

Lauwerijs, R.: *Carbon Monoxide, Environmental Pollution and Public Health.* Paper presented at the Conference on Health Effects of Carbon Monoxide Environmental Pollution, Luxembourg, 17th-19th December, 1973. Commission of the European Communities, 1974. Cited by Commins (1975).

Lauwerijs, R., Buchet, J.P., Roels, H., and Hubermont, G.: Placental transfer of lead, mercury, cadmium, and carbon monoxide in women. I. Comparison of the frequency distribution of the biological indices in maternal and umbilical cord blood. *Environ Res, 15:* 278-289, 1978.

Lawrence Berkeley Laboratory. *Instrumentation for Environmental Monitoring of Air. CO Monitoring Instrumentation.* University of California, Berkeley (Report LBL-1, Part 1C), 1973.

Lawther, P.J., and Apthorp, G.H.: A method for the determination of carbon monoxide in blood. *Br J Ind Med, 12*: 326-329, 1955.

Lawther, P.J., Commins, B.T., and Henderson, M.: Carbon monoxide in town air: an interim report. *Ann Occup Hyg, 5*: 241-246, 1962.

Leathart, G.L.: Hyperventilation in carbon monoxide poisoning. *Br Med J,* (ii): 511-512, 1962.

Ledwith, J.W.: Determining P_{50} in the presence of carboxyhemoglobin. *J Appl Physiol, 44:* 317-321, 1978.

Lee, K.T., Kit, O.A., and Jacob, E.: Determination of carboxyhemoglobin. *Mikrochim Acta, 2:* 657-663, 1975.

Le Groumellec, A., Duhamel, Y., and Simon, L.: Etude de l'affinité de l'hémoglobine pour l'O_2 par la méthode de la P_{50} chez les fumeurs. *Ouest Médical, 31:* 181-184, 1978.

Le Moan, M.G.: Semi-microdosage de l'oxyde de carbone sanguin. *Ann Pharm Fr, 10:* 269-276, 1952.

Lenfant, C., Torrance, J., English, E., et al.: Effect of altitude on oxygen binding by hemoglobin and on organic phosphate levels. *J Clin Invest, 47:* 2652-2656, 1968.

Levin, A.M., and Filatove, N.N.: Pollution of the inside air of gas-using accommodations with carbon monoxide. *Okhrana Truda I Okruzh Sredy, 1:* 62-66, 1977.

Levy, A.L., Lum, G., and Abeles, F.J.: Carbon monoxide in firemen before and after exposure to smoke. *Ann Clin Lab Sci, 6:* 455-458, 1976.

Levy, H.: II. Photochemistry of minor constituents in the troposphere. *Planet Space Sci, 21:* 575-591, 1973.

Lewey, F.H., and Drabkin, D.L.: Experimental chronic carbon monoxide poisoning of dogs. *Am J Med Sci, 208:* 502-511, 1944.

Lewin, L.: *Carbon Monoxide Poisoning: A Manual for Physicians, Engineers and Accident Investigators.* New York, Springer-Verlag, 1920, pp. 1-370.

Lewis, J., Baddeley, A.O., Bonham, K.G., and Lovett, D.: Traffic pollution and mental efficiency. *Nature, 225:* 95-97, 1970.

Lightfoot, N.F.: Chronic carbon monoxide exposure. *Proc R Soc Med, 65:* 16-17, 1972.

Lilienthal, J.L., and Fugitt, C.H.: The effect of low concentrations of COHb on the "altitude tolerance" of man. *Am J Physiol, 145:* 359-364, 1946.

Lilienthal, J.L., and Pine, M.B.: The effect of oxygen pressure on the uptake of carbon monoxide by man at sea level and at altitude. *Am J Physiol, 145:* 346-350, 1946.

Lilienthal, J.L., Riley, R.L., Prommel, D.D., and Franke, R.E.: The relationships between carbon monoxide, oxygen and hemoglobin in the blood of man at altitude. *Am J Physiol, 145:* 351-358, 1946.

Lily, R.E., Cole, PV., and Hawkins, L.H.: Spectrophotometric measurement of carboxyhemoglobin. An evaluation of the method of Commins and Lawther. *Br J Ind Med, 29:* 454-457, 1972.

Linch, A.L., and Pfaff, H.V.: Carbon monoxide—evaluation of exposure potential by personnel monitor surveys. *Am Ind Hyg Assoc J, 32:* 745-752, 1971.

Lindenberg, R., Preziosi, T., Levy, D., and Christensen: *Experimental Investigation of Carbon Monoxide Toxicity.* Paper presented at the Air Pollution Medical Research Conference, Los Angeles, California, December 4-7, 1961.

Linderholm, H.: A micromethod for the determination of carbon monoxide in blood. *Acta Physiol Scand, 64:* 372-376, 1965.

Linderholm, H., Sjöstrand, T. and Söderström, B.: A method for determination of low carbon monoxide concentration in blood. *Acta Physiol Scand, 66:* 1-8, 1966.

Lindgren, S.A.: A study of the effect of protracted occupational exposure to carbon monoxide. *Acta Med Scand, 356 (Suppl):* 5-135, 1961.

Lingen, M.L.: Effects of low concentrations of carbon monoxide on flicker fusion. *Nord Hyg Tidskr, 36:* 202-211, 1955.

Little, A.D.: *Energy Conservation in New Building Design. An Impact Assessment of ASHRAE Standard 90-75*. Washington, D.C., U.S. Govt. Print. Office (National Tech. Inf. Service PB-252 639), 1976, pp. 1-257.

Loeschke, H.H., and Gertz, K.H.: Einfluss des O_2—Druckes in der Einatmungsluft auf die auf die Atemtätigkeit des Menschen, geprüft unter Konstanthaltung des alveolaren CO_2—Druckes. *Pflügers Arch Ges Physiol, 267:* 460-477, 1958.

Logue, G.L., Rosse, W.F., Smith, W.T., et al.: Endogenous carbon monoxide production measured by gas phase analysis: An estimation of heme catabolic rate. *J Lab Clin Med, 77:* 867-876, 1971.

Loke, J., Farmer, W.C., Matthay, R., et al.: Carboxyhemoglobin levels in fire fighters. *Lung, 154:* 35-39, 1976.

Long, P.H.: The problem of smoking. *Med Times, 92:* 411-412, 1964.

Longo, L.D.: Carbon monoxide in the pregnant mother and fetus and its exchange across the placenta. *Ann NY Acad Sci, 174:* 313-341, 1970.

Longo, L.D.: Disorders of placental transfer. In Assali, N.S., and Brinkman, C.R. (Eds.): *Pathophysiology of Gestation*. New York, Academic Press, 1972, pp. 1-76.

Longo, L.D.: Carbon monoxide. Effects on oxygenation of the fetus in utero. *Science, 194:* 523-525, 1976.

Longo, L.D.: The biological effects of carbon monoxide on the pregnant woman, fetus and newborn infant. *Am J Obstet Gynecol, 129:* 69-103, 1977.

Longo, L.D., Power, G.G., and Forster, R.E.: Respiratory function of the placenta as determined with carbon monoxide in sheep and dogs. *J. Clin Invest, 46:* 812-828, 1967.

Longo, L.D., and Hill, E.P.: Carbon monoxide uptake and elimination in fetal and maternal sheep. *Am J Physiol, 232:* H324-330, 1977.

Longo, L.D., Wyatt, J.F., Hewitt, C.W., and Gilbert, R.D.: A comparison of circulatory responses to hypoxic hypoxia and carbon monoxide hypoxia in fetal blood flow and oxygenation. In Longo, L.D., and Reneau, D.D.: *Fetal and Newborn Cardiovascular Physiology: Fetal and Newborn Circulation. Vol 2*. New York, Garland STPM Press, 1978, pp. 259-287.

Lorang, P.: *Review of Past Studies Addressing the Potential Impact of Carbon Monoxide, Hydrocarbons and Nitrogen Oxide Emissions from Commercial Aircraft on Air Quality*. Report A.C. 78-03, 1978, pp. 1-73.

Loughbridge, L.W., Leader, L.P., and Bowen, D.A.L.: Acute renal failure due to muscle necrosis in carbon monoxide poisoning. *Lancet, 2:* 349-351, 1958.

Lown, B.: Sudden cardiac death: the major challenge confronting contemporary cardiology. *Am J Cardiol, 43:* 313-328, 1979.

Lucas, K.C.: National ambient air quality objectives promulgated. *Canada Gazette*, Part I, 3736-3739, 1972.

Luederitz, P.: Experiments on the effects of carbon monoxide from car exhaust fumes on the health of traffic police. *Z Gesamte Hyg, 17:* 645-646, 1971.

Lundh, B., Ohlin, P., and Westling, H.: Starka och Svaga Cigarretter — Med och Utan Filter. *Lakartidningen, 74:* 2356-2358, 1977.

Lundwig, G.D., and Blakemore, W.S.: Production of carbon monoxide by hemin oxidation. (Abstract). *J Clin Invest, 36:* 912, 1957.

Luomanmaki, K., and Coburn, R.F.: Effects of metabolism and distribution of carbon monoxide on blood and body stores. *Am J Physiol, 217:* 354-363, 1969.

Luria, S.M.: Visual masking and carbon monoxide toxicity. *Percept Mot Skills, 44:* 47-53, 1977.

Luria, S.M., and McKay, C.L.: Effects of low levels of carbon monoxide on vision of smokers and non-smokers. *Arch Environ Health, 34:* 38-44, 1979.

Lynch, S.R., and Moede, A.L.: Variation in the rate of endogenous carbon monoxide production in normal human beings. *J Lab Clin Med, 79:* 85-95, 1972.

MacMahon, B., Alpert, M., and Salber, E.J.: Infant weight and parental smoking habits. *Am J Epidemiol, 82:* 247-261, 1965.

MacMillan, D.E. Cited by DuBois (1969).

Maisels, M.J., Pathak, A., Nelson, N.M., et al.: Endogenous production of carbon monoxide in normal and erythroblastotic newborn infants. *J Clin Invest, 50:* 1-8, 1971.

Makishima, K., Keane, W.M., Vernose, G.V., and Snow, J.B.: Hearing loss of a central type secondary to carbon monoxide poisoning. *Trans Am Acad Ophthalmol Otolaryngol, 84:* 452-457, 1977.

Malenfant, A.L., Gambino, S.R., Waraska, A.J., and Roe, E.I.: Spectrophotometric determination of haemoglobin concentration and percent oxyhaemoglobin and carboxyhaemoglobin saturation. (Abstract). *Clin Chem, 14:* 789, 1968.

Malinow, M.R., McLaughlin, P., Dhindsa, D.S., et al.: Failure of carbon monoxide to induce myocardial infarction in cholesterol-fed Cynomolgus Monkeys *(Macaca fascicularis). Cardiovasc Res, 10:* 101-108, 1976.

Malleson, A.: *The Medical Run Around.* New York, Hart Publishing, 1973.

Malorny, G., Muller-Plathe, O., and Straub, M.: Disturbance of acid-base balance in acute carbon monoxide poisoning and effect of oxygen on carbon dioxide-oxygen mixtures (In German). *Arch Exp Pathol Pharmakol, 243:* 439-458, 1962.

Manning, F.A., and Feyerabend, C.: Cigarette smoking and fetal breathing movements. *Br J Obstet Gynecol, 83:* 262-270, 1976.

Manz, A., Zimmer, U., and Krusche, D.: CO-Gehalt des Blutes bei Ofenblockarbeitern einer modernen Kokereianlage. Zugleich ein Betrag zur Frage der CO — Belastung durch Tabakrauchen.*Zentralbl Arbeitsmed Arbeitsschutz Prophyl, 26:* 121-127, 1976.

Marriott, H.J.: Gases in artificial respiration. *Br Med J, 1:* 664-665, 1955.

Marriott, H.J.: Carbon monoxide poisoning. *Br Med J, 2:* 1591-1592, 1958.

Marshall, M.: Atherogenese, Experimentelle Untersuchungen zur Pathogenese oblitierender Arteriopathien am Miniaturschwein. *Munch Med Wschr, 121:* 1211-1218, 1979.

Marshall, R., Hess, H., and Staubesand, J.: Experimentelle Untersuchungen uber den Risikofaktor Rauchen. *Z Gefaesskrankheiten, 7:* 389-397, 1978.

Martin, A.E., and Oeter, D.: Environmental health aspects of human settlements. Copenhagen, World Health Organization Regional Office for Europe, 1978.

Mason, J.K.: *Aviation Accident Pathology.* London, Butterworth, 1962.

Mass, A.H.J., Hamelink, M., and de Leeuw, R.J.: An evaluation of the spectro-photometric determination of HbO_2, HbCO and Hb in blood with the CO-Oximeter IL 182. *Clin Chim Acta, 29:* 303-310, 1970.

Massaro, D.W., and Kahn, B.J.: Effects of central processing on auditory recognition. *J Exp Psychol, 97:* 51-58, 1973.

Master, K.: Carbon monoxide, atherosclerosis and natives of the highlands of New Guinea. *N Engl J Med, 294:* 556-557, 1976.

Mastromatteo, E.: Mortality in city firemen. I. A review. *Arch Ind Health, 20:* 1-7, 1959.

Mastromatteo, E.: Mortality in city firemen. II. A study of mortality in firemen of a city fire department. *Arch Ind Health, 20:* 227-233, 1959.

Maugh, T.H.: Carbon monoxide: natural sources dwarf man's output. *Science, 177:* 338-339, 1972.

Maugh, T.H.: Sulfuric acid from cars: a problem that never materialized. *Science, 198:* 280-284, 1977.

Mautner, L.S.: Muscle necrosis associated with carbon monoxide poisoning. *AMA Arch Pathol, 60:* 136-138, 1955.

Maxwell, J. Cigarette smoking and lung cancer. *Br Med J, 1:* 872-873, 1962.

McClatchie, E.A., Compher, A.B., and Williams, K.G.: A high specificity carbon monoxide analyzer. *Anal Instrum (Symposium), 10:* 67-69, 1972.

McConnell, J.C., McElroy, M.B., and Wofsy, S.C.: Natural sources of atmospheric CO. *Nature, 233:* 187-188, 1971.

McCormick, R.A., and Xintaras, C.: Variation of carbon monoxide concentrations as related to sampling interval, traffic and meteorological factors. *J Appl Meteorol, 1:* 237-243, 1962.

McCredie, R.M., and Jose, A.D.: Analysis of blood carbon monoxide and oxygen by gas chromatography. *J Appl Physiol, 22:* 863-866, 1967.

McCullough, J.D., Crane, R.A., and Beckman, A.O.: Determination of carbon monoxide in air by use of red mercuric oxide. *Anal Chem, 19:* 999-1002, 1947.

McFarland, R.A.: The psychological effects of oxygen deprivation (anoxemia) on human behavior. *Arch Psychol, 22:* 1-35, 1932.

McFarland, R.A.: Psycho-physiological studies at high altitude in the Andes. I. The effects of rapid ascents by aeroplane and train. *J Comp Psychol, 23:* 191-225, 1937a.

McFarland, R.A.: Psycho-physiological studies at high altitude in the Andes. II. Sensory and motor responses during acclimatization. *J Comp Psychol, 23:* 227-258, 1937b.

McFarland, R.A.: Experimental evidence of the relationship between aging and oxygen want: in search of a theory of aging. *Ergonomics, 6:* 339-366, 1963.

McFarland, R.A.: Low level exposure to carbon monoxide and driving performance. *Arch Environ Health, 28:* 355-359, 1973.

McFarland, R.A., and Halperin, M.H.: The relation between foveal visual acuity and illumination under reduced oxygen tension. *J Gen Physiol, 23:* 613-630, 1940.

McFarland, R.A., Roughton, F.J.W., Halperin, M.H., and Niven, J.J.: The effects of carbon monoxide and altitude on visual thresholds. *J Aviat Med, 15:* 381-394, 1944.

McIlvaine, P.M., Nelson, W.C., and Bartlett, D.: Temporal variation of carboxyhemoglobin concentrations. *Arch Environ Health, 19:* 83-91, 1969.

McMillan, D.E.: Interactions between carbon monoxide and d-amphetamine or pentobarbital on schedule controlled behavior. *Environ Res, 8:* 53-63, 1974.

McMullen, T.B.: Interpreting the eight-hour national ambient air quality standard for carbon monoxide. *J Air Pollut Control Assoc, 25:* 1009-1014, 1975.

McNall, P.E.: Practical methods of reducing airborne contaminants in interior spaces. *Arch Environ Health, 30:* 552-556, 1975.

Medical College of Wisconsin. *Exposure of Humans to Carbon Monoxide Combined with Ingestion of Ethyl Alcohol and the Comparison of Human Performance when Exposed to Varying Periods of Time to Carbon Monoxide.* Report MCOW-ENVM-CO-74-2. Medical College of Wisconsin, 1974.

Medical Research Council. Carbon monoxide poisoning: use of CO_2-O_2 mixture. *Br Med J,* (ii): 1408-1409, 1958.

Mercke, C., Cavallin-Stähl, E., and Lundh, B.: Diurnal variation in endogenous production of carbon monoxide. *Acta Med Scand, 198:* 161-164, 1975.

Middleton, V., Van Poznak, A., Smith, S.M., and Artusio, J.F.: Carbon monoxide accumulation in closed circle anesthesia systems. *Anesthesiology, 26:* 715-719, 1965.

Mikulka, P., O'Donnell, R., Heinig, P., and Theodore, J.: The effect of carbon monoxide on human performance. *Ann NY Acad Sci, 174:* 409-420, 1970.

Miles, S.: *Underwater Medicine.* London, Staples Press, 1962.

Millar, R.A., and Gregory, I.C.: Reduced oxygen content on equilibrated fresh heparinized and ACD-stored blood from cigarette smokers. *Br J Anaesth, 44:* 1015-1019, 1972.

Miller, G.A., and Taylor, W.G.: The perception of repeated bursts of noise. *J Acoust Soc Am, 20:* 171-182, 1948.

Miller, V.L., and Engel, R.R.: In-vivo metabolism of di-halomethanes to carbon monoxide. *Pharmacologist, 15:* 190, 1973.

Millikan, G.A.: Muscle haemoglobin. *Physiol Rev, 19:* 503-520, 1939.

Mills, E., and Edwards, McI.W.: Stimulation of aortic and carotid chemoreceptors during carbon monoxide inhalation. *J Appl Physiol, 25:* 494-502, 1968.

Ministry of Labour. *Carbon Monoxide Poisoning: Causes and Prevention.* London, HMSO (Safety, Health, and Welfare, New Series#29), 1965.

Miranda, J.M., Konopinski, V.J., and Larsen, R.I.: Carbon monoxide control in a high highway tunnel. *Arch Environ Health, 15:* 16-25, 1967.

Mitchell, R.S., Judson, F.N., Moulding, T.S., et al.: Health effects of urban air pollution. Special consideration of areas at 1,500m and above. *J Am Med Assoc, 242:* 1163-1167, 1979.

Montgomery, M.R., and Rubin, R.J.: The effect of carbon monoxide inhalation in "In Vivo" drug metabolism in the rat. *J Pharmacol Exp Ther, 179:* 465-473, 1971.

Morel, S., and Szemberg, T.: Air pollution in Ladle Bay. *Tecknika Sanitarna, 45:* 353-355, 1971.

Moriyama, J.M., Drueger, D.E., and Stamler, J.: *Cardiovascular Diseases in the United States.* Cambridge, Massachusetts, Harvard University Press, 1971.

Morrison, J.H.: *Carbon Monoxide Determination Feasibility Study. Reports on the Epidemiology and Surveillance of Injuries.* Washington, D.C., U.S. Govt. Print. Office, (DHEW Publication, HSMHA, BCEM. FY 72-RC), 1972.

Morrison, J.H.: *Analysis of Carbon Monoxide Death Certificates for 1967.* Washington, D.C., U.S. Govt. Print. Office (DHEW Publication BCEMI), 1974.

Morrow, P.E.: An evaluation of recent NO_x toxicity data and an attempt to derive an ambient air standard for NO_x by established toxicological procedures. *Environ Res, 10:* 92-112, 1975.

Moureau, H.: Carbon monoxide as a test for air pollution in Paris due to motor vehicle traffic. *Proc R Soc Med, 57:* 1015-1020, 1964.

Moureau, H., Chovin, P., Truffert, L., and Lebbe, J.: Nouvelle microméthode pour la détermination rapide et précise de l'oxycarbonémie, par absorption sélective dans l'infrarouge. *Arch Mal Prof, 18:* 116-124, 1957.

Mulhausen, R.O., Åstrup, P., and Mellemgaard, K.: Oxygen affinity and acid-base status of human blood during exposure to hypoxia and carbon monoxide. *Scand J Clin Lab Invest, 22, Suppl. 103:* 9-15, 1968.

Müller, R.H.: A supersensitive gas detector permits accurate detection of toxic or combustible gases in extremely low concentrations. *Anal Chem, 26:* 39A-42A, 1954.

Mumpower, R.C., Lewis, J.S., and Touey, G.P.: Determination of carbon monoxide in cigarette smoke by gas chromatography. *Tobacco Science, 6:* 142-145, 1962.

Munro, L.A.: *Chemistry in Engineering.* Englewood Cliffs, New Jersey, Prentice-Hall Inc., 1964.

Murphy, S.D.: A review of effects on animals of exposure to auto exhaust and some of its components. *J Air Pollut Control Assoc, 14:* 303-307, 1964.

Myrhe, L.: Rate of CO elimination in resting man. *Fed Proc, 33:* 44OP, 1974.

Myronuk, D.J.: Augmented ingestion of carbon monoxide and sulfur oxides by occupants of vehicles while idling in drive-up facility lines. *Water Air Soil Pollut, 7:* 203-213, 1977.

Nakajima, T., Hattori, S., Tateisui, R., and Horai, T.: Morphological changes in the bronchial alveolar system of mice following continuous exposure to low concentrations of nitrogen dioxide and carbon monoxide. *J Jap Soc Chest Dis, 10:* 16-22, 1972.

Nasmith, G.G., and Graham, D.A.L.: The haematology of carbon monoxide poisoning. *J Physiol, 35:* 32-52, 1906.

National Academy of Sciences. *Guides for Short-Term Exposure of the Public to Air Pollutants. VI. Guide for Carbon Monoxide.* Washington, D.C., U.S. Govt. Print. Office, 1972.

National Academy of Sciences. *Report for Environmental Protection Agency,* Washington, D.C., U.S. Govt. Print. Office (PB 224-862), 1972.

National Academy of Sciences. *Carbon Monoxide.* Washington, D.C., U.S. Govt. Print. Office, 1977.

National Academy of Sciences, and National Academy of Engineering. *Air Quality and Automobile Emission Control Vol. 2. Health Effects of Air Pollutants. Vol. 4. The Costs and*

Benefits of Automobile Emission Control. (Committee on Public Works, US Senate 93-24). Washington, D.C., U.S. Govt. Print. Office, 1974.

National Center for Disease Control. *Occupational Exposure to CO in Selected Rural Work Environments.* Atlanta, Georgia, 1973.

National Center for Disease Control. *State Legislation on Smoking and Health.* Washington, D.C., U.S. Govt. Print. Office (DHEW Publication CDC-79-8331), 1978.

National Clearing House. *State Legislation on Smoking and Health.* Washington, D.C., U.S. Govt. Print. Office (DHEW Publication), 1976, pp. 1-73.

National Clearing House. *State Legislation on Smoking and Health.* Washington, D.C., U.S. Govt. Print. Office (DHEW Publication CDC-78-8331), 1977.

National Institute for Occupational Safety and Health. *Occupational Exposure to Carbon Monoxide: Criteria for a Recommended Standard.* Washington, D.C., U.S. Govt. Print. Office (DHEW Publication), 1972.

National Safety Council. *Carbon Monoxide.* Data Sheet 415 (Revision A), 1969.

Necheles, T.F., Rai, S.U., and Valaes, T.: The role of hemolysis in neonatal hyperbilirubinemia as reflected in carboxyhemoglobin levels. *Acta Paediatr Scand, 65*: 361-367, 1976.

Neil, E., and Joels, N.: The carotid glomus sensory mechanism. In Cunningham, D.J., and Lloyd, B.B. (Eds.): *Regulation of Human Respiration.* Oxford, Blackwell, 1963, pp. 163-171.

Neuberger, M.B.: *Smoke Screen: Tobacco and Public Welfare.* Englewood Cliffs, New Jersey, Prentice Hall, 1963.

Newby, D.W., Roberts, R.J., and Bhatnagar, R.K.: Carbon monoxide hypoxia induced effects on catecholamines in the mature and developing rat brain. *J Pharmacol Exp Ther, 206*: 61-68, 1978.

New England Journal of Medicine: Editorial. *N Eng J Med, 261*: 1248-1249, 1959.

New Mexico State Department of Health. *Birthweight and Altitude*] Albuquerque, New Mexico, New Mexico Department of Health, 1975.

Newsome, J.R., and Keith, C.H.: Variation of the gas phase composition within a burning cigarette. *Tobacco Science, 9*: 65-69, 1965.

Nicholson, S.E.: A pollution model for street-level air. *Atmos Environ, 9*: 19-31, 1975.

Niden, A.H., and Schultz, H.: The ultra-structural effects of carbon monoxide inhalation on the rat lung. *Virchows Arch (Pathol Anat), 339*: 283-292, 1965.

Noe, L.J., Eisert, W.G., and Rentzepis, P.M.: Pico second photo dissociation and subsequent recombination processes in carboxyhemoglobin, carboxymyoglobin and oxymyoglobin. *Biophys J, 24*: 379-381, 1978.

Norman, J.N., Douglas, T.A., and Smith, G.: Respiratory and metabolic changes during carbon monoxide poisoning. *J Appl Physiol, 21*: 848-852, 1966.

Norman, J.N., and Ledingham, I. McA.: Carbon monoxide poisoning: investigations and treatment. *Prog Brain Res, 24*: 101-122, 1967.

Norton, S., and Culver, B.: A Golgi analysis of caudate neurons in rats exposed to carbon monoxide. *Brain Res. 132*: 455-465, 1977.

Notov, A.: Acute poisonings with CO, SO_2 and nitric gases in mines (Bulgarian). *Sb Trudove VMI Plovidiv, 60:* 404-411, 1959.

O'Donnell, R.D., Mikulka, P., Heinigs, P., and Theodore, J.: *Effects of Short-Term Low Level Carbon Monoxide Exposure on Human Performance.* U.S. Aerospace Medical Research Laboratory (Report AMRL-TR-70-37), 1970.

O'Donnell, R.D., Chikos, P., and Theodore, J.: Effect of carbon monoxide exposure on human sleep and psychomotor performance. *J Appl Physiol, 31*: 513-518, 1971.

O'Donnell, R.D., Mikulka, P., Heinig, P., and Theodore, J.: Low level carbon monoxide exposure and human psychomotor performance. *Toxicol Appl Pharmacol, 18*: 593-602, 1971.

Okada, Y., Tyuma, I., Ueda, Y., and Sugimoto, T.: Effect of carbon monoxide on equilibrium between oxygen and hemoglobin. *Am J Physiol, 230:* 471-475, 1976.

Ontario Medical Association: *Exercise Prescription. A Physician's Guide to Assessment of Patients who Plan to Exercise.* Toronto, Ontario Ministry of Health, 1976.

Ontario Ministry of the Environment: *Report on Carbon Monoxide Sampling in Downtown Toronto - Summer, 1972.* October, 1972.

Opitz, E.: Increased vascularization of the tissue due to acclimatization to high altitude and its significance for oxygen transport. *Exp Med Surg, 9*:389-403, 1951.

Osborne, J.S., Adamek, S., and Hobbs, M.E.: Some components of gas phase of cigarette smoke. *Anal Chem, 28*: 211-215, 1956.

Occupational Safety and Health Administration: *U.S. Department of Labor, Occupational Safety and Health Standards. Amendments. August 13, 1971. Federal Register, 36*: 15102, 1971.

Oski, F.A., and Altman, A.A.: Carboxyhemoglobin levels in hemolytic disease of the newborn. *J Pediatr, 61*: 709-713, 1962.

Otis, A.B.: The physiology of carbon monoxide poisoning and evidence for acclimatization. *Ann NY Acad Sci, 174:* 242-245, 1970.

Ott, W.R., and Eliassen, R.: A survey technique for determining the representativeness of urban air monitoring stations with respect to carbon monoxide. *J Air Pollut Control Assoc, 23*: 685-690, 1973.

Ott, W.R., and Mage, D.T.: Interpreting urban carbon monoxide concentrations by means of a computerized blood COHb model. *J Air Pollut Control Assoc, 28:* 911-916, 1978.

Otto, D.A., Benignus, V.A., and Prah, J.D.: Carbon monoxide and human time discrimination: failure to replicate Beard-Wertheim experiment. *Aviat Space Environ Med, 50*: 40-43, 1979.

Ouellet, V.: *The Clean Air Act. Compilation of Regulations and Guidelines.* Ottawa, Environment Canada Report (EPS 1-AP-78-2), 1978.

Pace, N., Consolazio, W.V., White, W.A., and Behnke, A.R.: Formulation of the principal factors affecting the rate of uptake of carbon monoxide by man. *Am J Physiol, 147*: 352-359, 1946.

Pace, N., Stajman, E., and Walker, E.L.: Acceleration of carbon monoxide elimination in man by high pressure oxygen. *Science, III:* 2893, 652-654, 1950.

Paintal, A.S.: Mechanism of stimulation of aortic chemoreceptors by natural stimuli and chemical substances. *J Physiol, 189*: 63-84, 1967.

Palanos, P.N.: A practical design for an ambient carbon monoxide mercury replacement analyzer. *Anal Instrum (Symposium), 10*: 117-125, 1972.

Palek, J., Brabec, V., Mircevová, L., et al.: Red cell carbohydrate metabolism in primary refractory anemias. *Clin Chim Acta, 18*: 39-45, 1967.

Pandolfelli, E.R., Bonaventura, C., and Bonaventura, J.: Light jump perturbation of carbon monoxide binding by various heme proteins. *Biophys J, 24*: 257, 1978.

Pankow, D., Ponsold, W., and Fritz, H.: Combined effects of carbon monoxide and ethanol on the effects of leucine aminopeptidase and glutamic-pyruvic transaminase in the plasma of rats. *Arch Toxikol, 32*: 331-340, 1974.

Pankow, D., Glatzel, W., Tietze, K., and Ponsold, W.: Motor nerve conduction velocity after carbon monoxide or m-dinitrobenzene poisoning, following elimination of the poisons. *Arch Toxicol, 34*: 325-330, 1975.

Paprowski, D.M., and Walker, J.R.: *Air Quality in Canadian Urban Areas. Discussion Paper #18.* Ottawa, Ontario, Economic Council of Canada, 1974.

Parving, H.H.: The effect of hypoxia and carbon monoxide exposure on plasma volume and capillary permeability to albumin. *Scand J Clin Lab Invest, 30*: 49-56, 1972.

Paul, K.G., and Theorell, H.: A colorimetrical carbon-monoxide-hemoglobin method of determination for clinical use. *Acta Physiol Scand, 4*: 285-292, 1942.

Pauli, H.G., Truniger, B., Larsen, J.K., and Mulhausen, R.O.: Renal function during prolonged exposure to hypoxia and carbon monoxide. I. Glomerular filtration and plasma flow. *Scand J Clin Lab Invest, 22, Suppl. 103*: 55-60, 1968.

Pauli, H.G., Truniger, B., Larsen, J.K., and Mulhausen, R.O.: Renal function during prolonged exposure to hypoxia and carbon monoxide. II. Electrolyte handling. *Scand J Clin Lab Invest, 22, Suppl. 103*: 61-67, 1968.

Pauling, L.: Resonant energy of carbon monoxide. In *The Nature of the Chemical Bond and the Structure of Molecules and Crystals. An Introduction to Modern Structural Chemistry,* 3rd ed. Ithaca, N.Y., Cornell University Press, 1960, pp. 194-195.

Penney, D., Benjamin, M., and Dunham, E.: Effect of carbon monoxide on cardiac weight as compared with altitude effects. *J Appl Physiol, 37*: 80-84, 1974.

Pennsylvania Department of Health. *Short-Term Limits for Exposure to Airborne Contaminants: A Documentation.* Harrisburg, Pennsylvania, Pennsylvania Department of Health, 1972.

Permutt, S., and Farhi, L.: Tissue hypoxia and carbon monoxide. In DuBois, A.B. (Ed.): *Effects of Chronic Exposure to Low Levels of Carbon Monoxide on Human Health, Behavior and Performance.* Washington, D.C., National Academy of Sciences and National Academy of Engineering, 1969, pp. 18-24.

Perutz, M.F.: Stereochemistry of cooperative effects in haemoglobin. *Nature, 228:* 726-734, 1970.

Petajan, J.H.: Survival response during fire exposure. In *Physiological and Toxicological Aspects of Combustion Products.* NAS/NRC International Symposium, Salt Lake City, 1976, pp.105-118.

Petajan, J.H., Packham, S.C., Frens, D.B., and Dinger, B.G.: Sequelae of carbon monoxide-induced hypoxia in the rat. *Arch Neurol, 33*: 152-157, 1976.

Peterson, J.E.: Post-exposure relationship of carbon monoxide in blood and expired air. *Arch Environ Health, 21*: 172-173, 1970.

Peterson, J.E., and Stewart, R.D.: Absorption and elimination of carbon monoxide by inactive young men. *Arch Environ Health, 21*: 165-171, 1970.

Peterson, J.E., and Stewart, R.D.: Human absorption of carbon monoxide from high concentrations in air. *Am Ind Hyg Assoc J, 31:* 293-297, 1972.

Peterson, J.E., and Stewart, R.D.: Predicting the carboxyhaemoglobin levels resulting from carbon monoxide exposures. *J Appl Physiol, 39:* 633-638, 1975.

Petri, H.: *Die Chronische Kohlenoxydvergiftung. Leipzig, Johann Ambrosius Barth,* 1953.

Petrov, P.: Chronic CO intoxications in workers engaged in loading and disloading processes. (In Bulgarian). *Transporti Medicenski Vesti, 4*: 24-27, 1968.

Pfaffinger, I., Marshall, M., Hess, H., and Staubesand, J.: Die akute Wirkung von Zigarettenrauch oder Kohlenmonoxidinhalation auf die Arterienwand beim Miniaturschwein. *Verh Dtsch Ges f Inn Med, 22*: 887-890, 1979.

Pfrender, R.E.: Chronic carbon monoxide poisoning. A critical resumé. *Ind Med Surg, 31*: 99-103, 1962.

Phair, J.J., Carey, G.C.R., and Shephard, R.J.: *Measuring Human Reactions to Air Pollution.* Philadelphia, Pennsylvania, Monograph *4*, Franklin Institute, 1958.

Phair, J.J., and Sterling, T.: Epidemiological methods and community air pollution. *Arch Environ Health, 3*: 267-275, 1961.

Pham, Q.T., Mur, J.M., Beigbeder, R., et al.: Long-term effects of low concentrations of noxious gases on the respiratory system: results of an epidemiological study among the iron mines workers in Lorraine (France). *Ref Epidémiol Santé Publique, 25*: 255-273, 1978.

Pierce, J.O., and Collins, R.J.: Calibration of an infrared analyzer for continuous measurement of carbon monoxide. *Am Ind Hyg Assoc J, 32:* 457-462, 1971.

Pilaar, W.M.M.: Determination of carbon monoxide in blood. *J Biol Chem, 83*: 43-50, 1929.

Pirnay, R., Dujardin, J., Deroanne, R., and Petit, J.M.: Muscular exercise during intoxication by carbon monoxide. *J Appl Physiol, 31*: 573-575, 1971.

Pitts, G.C., and Pace, N.: The effect of blood carboxyhemoglobin concentration on hypoxia tolerance. *Am J Physiol, 148*: 139-151, 1947.

Place, M.: For each his own—the self-rescuer apparatus against CO. *Occup Health (Lond), 22:* 107-111, 1970.

Plomp, R.: Rate of decay of auditory sensation. *J Acoust Soc Am, 36*: 277-282, 1964.

Polis, R.D., Berger, L.B., and Schrenk, H.H.: *Colorimetric Determination of Low Concentration of CO by use of a Pd Cl_2-phosphomolybdic Acid-Acetone Reagent.* Washington, D.C., U.S. Govt. Print. Office, (U.S. Bureau of Mines, Report of Investigations *3785*), 1944, p. 13.

Pollard, M.R., and Brennan, J.T.: Disease prevention and health promotion initiatives: some legal considerations. *Health Educ Monogr, 6*: 211-222, 1978.

Ponsold, W., Pankow, D., Grimm, I., et al.: Synergistic effect of carbon monoxide and benzene in rats. *Z Gesamte Hyg, 24*: 508-513, 1978.

Porter, K., and Volman, D.H.: Flame ionization detection of carbon monoxide for gas chromatographic analysis. *Anal Chem, 34*: 748-749, 1962.

Power, G.G., and Longo, L.D.: Fetal circulation times and their implications for tissue oxygenation. *Gynecol Invest, 6*: 342-355, 1975.

Preziosi, T.J., Lindenberg, R., Leog, D., and Christensen, M.: An experimental investigation in animals of the functional and morphologic effects of single and repeated exposures to high and low concentrations of carbon monoxide. *Ann NY Acad Sci, 174*: 369-384, 1970.

Prigge, E., and Hochrainer, D.: Effects of carbon monoxide inhalation on erythropoiesis and cardiac hypertrophy in fetal rats. *Toxicol Appl Pharmocol, 42*: 225-228, 1977.

Pugh, L.G.C.E.: Carbon monoxide hazard in Antarctica. *Br Med J,* (i): 192-196, 1959.

Putz, V.R., Johnson, B.L., and Setzer, J.V.: *Effects of CO on Vigilance Performance. Effects of Low Level Carbon Monoxide on Divided Attention, Pitch Discrimination and the Auditory Evoked Potential.* Bethesda, Maryland, National Institute for Occupational Safety and Health (DHEW Publication NIOSH 77-124), 1976, pp. 1-58.

Putz, V.R., Johnson, B.L., and Setzer, J.V.: A comparative study of the effects of carbon monoxide and methylene chloride on human performance. *J Environ Pathol Toxicol, 2*: 97-112, 1979.

Radford, E.P.: *Survey of Carbon Monoxide in East Baltimore Housing–A Preliminary Report.* Baltimore, Johns Hopkins University, 1973.

Radford, E.P.: Letter to the editor. *JAMA, 232*: 487, 1975.

Radford, E.P.: Carbon Monoxide and human health. *J Occup Med, 18*: 310-315, 1976.

Radford, E.P., and Kresin, E.: Red cell diphosphoglycerate after exposure to carbon monoxide. (Abstract.) *Fed Proc, 32*: 350, 1973.

Radford, E.P., and Rodkey, F.L.: Methods for measurement of carbon monoxide. In DuBois, A.B. (Chairman): *Effects of Carbon Monoxide on Man's Health in his Home Environment.* Washington, D.C., U.S. Govt. Print. Office (DHEW Publication), 1973.

Radford, E.P., and Levine, M.: Occupational exposure to carbon monoxide in Baltimore firefighters. *J Occup Med, 76*: 628-632, 1976.

Raeder, E.A., Burckhardt, D., Perruchoud, A., et al.: Effects of smoking and inhalation of carbon monoxide on systolic time intervals and blood pressure. Differences between two types of cigarettes and a cigar. *Chest, 75*: 136-140, 1979.

Ramieri, A., Jatlow, P., and Seligson, D.: New method for rapid determination of carboxyhemoglobin by use of double wavelength spectrophotometry. *Clin Chem, 20*: 278-281, 1974.

Ramsey, J.M.: Concentrations of carbon monoxide at traffic intersections in Dayton, Ohio. *Arch Environ Health, 13*: 44-46, 1966.

Ramsey, J.M.: Carboxyhemoglobinemia in parking garage employees. *Arch Environ Health, 15*: 580-583, 1967.

Ramsey, J.M.: Oxygen reduction and reaction time in hypoxic and normal drivers. *Arch Environ Health, 20*: 597-601, 1970.

Ramsey, J.M.: Carbon monoxide, tissue hypoxia and sensory psychomotor response in hypoxemia. *Clin Sci, 42*: 619-625, 1972.

Ramsey, J.M.: Effects of single exposures of carbon monoxide on sensory and psychomotor response. *Am Ind Hyg Assoc J, 34*: 212-216, 1973.

Randel, H.W.: *Aerospace Medicine*, 2nd ed. Baltimore; Williams and Wilkins, 1971.

Ratney, R.S., Wegman, D.H., and Elkins, H.B.: In vivo conversion of methylene chloride to carbon monoxide. *Arch Environ Health, 28*: 223-226, 1974.

Rauch Verbot bei Fortbildungsveranstaltungen. *Med Monatsschr, 29*: 141-142, 1975.

Raven, P.B., Drinkwater, B.L., Ruhling, R.O., et al.: Effect of carbon monoxide and peroxyacetyl nitrate on man's maximal aerobic capacity. *J Appl Physiol, 36*: 288-293, 1974.

Ray, A.M., and Rockwell, T.H.: An exploratory study of automobile driving performance under the influence of low levels of carboxyhemoglobin. *Ann NY Acad Sci, 174*: 396-407, 1970.

Rees, P.J., Chilvers, C., and Clark, T.J.H.: Evaluation of methods used to estimate inhaled dose of carbon monoxide. *Thorax, 35*: 47-51, 1980.

Remick, R.A., and Miles, J.E.: Carbon monoxide poisoning: neurologic and psychiatric sequelae. *Can Med Assoc J, 117*: 654, 657, 1977.

Reynafarje, B.: Myoglobin content and enzymatic activity of muscle and altitude adaptation. *J Appl Physiol, 17*: 301-305, 1962.

Rice, E.W.: Improved photometric determination of carbon monoxide by means of palladious chloride. *Arch Ind Hyg Occup Med, 6*: 487-490, 1952.

Richardson, N.A., and Middleton, W.C.: Evaluation of filters for removing irritants from polluted air. *ASHRAE Trans, 65*: 401-416, 1959.

Ringold, A., Goldsmith, J.R., Helwig, H.L., et al.: Estimating recent carbon monoxide exposure. A rapid method. *Arch Environ Health, 5*: 308-318, 1962.

Rink, R.D.: The acute effects of nicotine, tobacco smoke and carbon monoxide on myocardial oxygen tension in the anaesthetized cat. *Br J Pharmacol, 62*: 591-597, 1978.

Rispler, X., and Gilbert, C.: *Report of Carbon Monoxide Survey, Toronto International Airport*. Ottawa Occupational Health Division, Department of National Health and Welfare, 1970.

Roach, S.A.: A more rational basis for air sampling programs. *Am Ind Hyg Assoc J, 27:* 1-12, 1966.

Roach, S.A.: Industrial ventilation. In Schilling, R.S.F. (Ed.): *Occupational Health Practice*. Ed: R.S.F. Schilling. London, Butterworths, 1973, pp. 339-359.

Robbins, R.C., Borg, K.M., and Robinson, E.: Carbon monoxide in the atmosphere. *J Air Pollut Control Assoc, 18:* 106-110, 1968.

Robinson, E., and Robbins, R.C.: Atmospheric background concentrations of carbon monoxide. *Ann NY Acad Sci, 174:* 89-95, 1970.

Robinson, J.C., and Forbes, W.F.: The role of carbon monoxide in cigarette smoking. I. Carbon monoxide yield from cigarettes. *Arch Environ Health, 30:* 425-434, 1975.

Robinson, S.A., and Wolfe, S.M.: *Smoking: Its Adverse Effects on Airplane Pilot Performance*. Washington, D.C., Health Research Group, 1976, pp. 1-50.

Rockwell, T.H., and Ray, A.M.: *Subacute Carbon Monoxide Poisoning and Driving Performance: A Selected Review of the Literature and Discussion*. Chicago, Illinois, National Safety Council, 1967.

Rockwell, T.J., and Weir, F.W.: *The interactive effects of carbon monoxide and alcohol on driving skills*. Ohio State University Research Foundation (RF Project *332*), 1975, pp. 1-112.

Rode, A.: *Acute and Chronic Effects of Smoking on Fitness*. Master's thesis, School of

Hygiene, University of Toronto, 1970.

Rode, A., and Shephard, R.J.: Smoking withdrawal and changes of cardiorespiratory fitness. *Am Rev Respir Dis, 104:* 933-935, 1971.

Rode, A., Ross, R., and Shephard, R.J.: Personality and changes in the cardiorespiratory fitness of participants in a smoking withdrawal program. *Arch Environ Health, 24:* 27-36, 1972.

Rode, A., and Shephard, R.J.: Pulmonary function of Canadian Eskimos. *Scand J Respir Dis, 54:* 191-205, 1973.

Rodgers, S.J.: *Analysis of Noncoal Mine Atmospheres: Toxic Fumes from Explosives.* Washington, D.C., U.S. Govt. Print. Office, (U.S. Bureau of Mines, Department of Commerce PB 262-510), May 5, 1976.

Rodkey, F.L.: Kinetic aspects of cyanmethemoglobin formation from carboxyhemoglobin. *Clin Chem, 13:* 2-5, 1967.

Rodkey, F.L.: The measurement of carbon monoxide in biological fluids. In Sunderman, F.W., and Sunderman, F.W., Jr. (Eds.): *Laboratory Diagnosis of Diseases Caused by Toxic Agents.* St. Louis, Missouri, W.H. Green Inc., 1970, pp. 278-285.

Rodkey, F.L., O'Neal, J.D., and Collison, H.A.: Oxygen and carbon monoxide equilibria of human adult hemoglobin at atmospheric and elevated pressure. *Blood, 33:* 57-65, 1969.

Rodkey, F.L., O'Neal, J.D., Collison, H.A., and Uddin, D.E.: Relative affinity of Hemoglobin S and Hemoglobin A for carbon monoxide and oxygen. *Clin Chem, 20:* 83-84, 1974.

Rodkey, F.L., Raymond, L.W., Collison, H.A., and O'Neal, J.D.: Changes in blood carboxyhemoglobin during simulated saturation diving to 50 ATA. *Undersea Biomed Res, 1:* 197-201, 1974.

Rogowsky, M., Van der Putten, M., and Baudoux, A.: Etude critique de la formation de Carboxyhémoglobine (HbCO) chez les soudeurs travaillant en atmosphere à faible concentration d'oxyde de carbone (CO). *Arch Mal Prof, 39:* 151-161, 1978.

Rondia, D.: Abaissement de l'activité de la benzopyrène—hydroxylase hépatique in vivo après l'inhalation d'oxyde de carbone. *CR Acad Sci, (Paris), 271:* 617-619, 1970.

Root, W.S.: Carbon monoxide. In Fenn, W.O., and Rahn, H. (Eds.): *Handbook of Physiology. Section 3. Respiration.* Washington, D.C., American Physiological Society, 1965, pp. 1087-1098.

Rose, A.H., and Smith, R.: A direct measurement technique for automobile exhaust. *Arch Environ Health, 5:* 609-615, 1962.

Rose, A.H., Smith, R., McMichael, W.F., and Kruse, R.E.: Comparison of auto exhaust emissions in two major cities. *J Air Pollut Control Assoc, 15:* 362-366, 1965.

Rosenman, K.D.: Cardiovascular disease and environmental exposure. *Br J Ind Med, 36:* 85-97, 1979.

Ross, J.C.: Trace contaminants. In *Physiology in the Space Environment Vol 2. Respiration.* Washington, D.C., U.S. Govt. Print. Office (Publication 1485B), 1967, pp. 113-122.

Rossi-Fanelli, A., and Antonini, E.: Studies on the oxygen and carbon monoxide equilibria on human myoglobin. *Arch Biochem, 77:* 478-492, 1958.

Roughton, F.J.W.: The chemistry of respiration. In Cunningham, D.J.C., and Lloyd, B.B. (Eds.): *The Regulation of Human Respiration.* Oxford, Blackwell, 1963, pp. 33-44.

Roughton, F.J.W.: Transport of oxygen and carbon dioxide. In Fenn, W.O., and Rahn, H. (Eds.): *Handbook of Physiology. Section 3, Respiration I.* Washington, D.C., American Physiological Society, 1964, pp. 767-825.

Roughton, F.J.W.: The equilibrium of carbon monoxide with human hemoglobin in whole blood. *Ann NY Acad Sci, 174*, 177-188, 1970.

Roughton, F.J.W., and Darling, R.C.: The effect of carbon monoxide on the oxyhemoglobin dissociation curve. *Am J Physiol, 141:* 17-31, 1944.

Roughton, F.J.W., and Root, W.S.: The fate of CO in the body during recovery from mild carbon monoxide poisoning in man. *Am J Physiol, 145:* 239-252, 1945a.

Roughton, F.J.W., and Root, W.S.: The estimation of small amounts of carbon monoxide in blood. *J Biol Chem, 160:* 123-133, 1945b.

Rubino, G.F., Pettinati, L., and Capellaro, F.: La odorizzazione del gas illuminante quale mezzo di prevenzione dell' ossicarbonismo acuto. In *Ergonomics and Physical Environmental Factors.* Geneva, International Labour Organization, 1970, pp. 232-233.

Rummo, N., and Sarlanis, K.: The effect of carbon monoxide on several measures of vigilance in a simulated driving task. *J Safety Res, 6:* 126-130, 1974.

Russell, M.A.H., Cole, P.V., and Brown, E.: Absorption by non-smokers of carbon monoxide from room air polluted by tobacco smoke. *Lancet,* (i): 576-579, 1973.

Russell, M.A.H., and Cole, P.V.: Passive smoking. Absorption by non-smokers of carbon monoxide from room air polluted by tobacco smoke. *Postgrad Med J, 49:* 688-692, 1973.

Russell, M.A.H., Wilson, C., Cole, P.V., et al.: Comparison of increases in carboxyhaemoglobin after smoking "extra mild" and "non-mild" cigarettes. *Lancet* (ii): 687-690, 1973.

Russell, M.A.H., Wilson, C., Patel, U.A., et al.: Comparison of effect on tobacco consumption and carbon monoxide absorption of changing to high and low nicotine cigarettes. *Br Med J,* (iv): 512-516, 1973.

Russell, M.A.H., and Feyerabend, C.: Blood and urinary nicotine in non-smokers. *Lancet,* (i): 179-181, 1975.

Ryazanov, V.A.: Sensory physiology as a basis for air quality standards. *Arch Environ Health, 5:* 480-494, 1962.

Sairo, O., and El-Attar, A.: Effect of carbon monoxide on the whole blood fibrinolytic activity. *Ind Med Surg, 37:* 774-777, 1968.

Sammons, J.H., and Coleman, R.L.: Firefighters' occupational exposure to carbon monoxide. *J Occup Med, 16:* 543-546, 1974.

Santiago, T.V., and Edelman, N.H.: Mechanism of the ventilatory response to carbon monoxide. *J Clin Invest, 57:* 977-986, 1976.

Sarma, S.J.M., Tillmanns, H., Ikeda, S., and Bing, R.J.: The effect of carbon monoxide on lipid metabolism of human coronary arteries. *Atherosclerosis, 22:* 193-195, 1975.

Sawicki, C.A., and Gibson, Q.H.: The relationship between carbon monoxide binding and the conformational change of hemoglobin. *Biophys J, 24:* 21-32, 1978.

Sawicki, C.A., and Gibson, Q.H.: Photochemical method for rapid and precise determination of carbon monoxide levels in blood. *Anal Biochem, 94:* 440-449, 1979.

Sayers, R.R., Yout, W.P., Levy, E., and Fulton, W.B.: *Effect of Daily Exposure of Several Hours to Small Amounts of Automobile Exhaust Gas.* Washington, D.C., U.S. Govt. Print. Office (Public Health Bulletin #186), 1929.

Sayers, R.R., and Davenport, S.J.: *Review of Carbon Monoxide Poisoning.* Washington, D.C., U.S. Govt. Print. Office (Public Health Bulletin #195), 1936.

Schievelbein, H.: Evaluation of the role of carbon monoxide and nicotine in the pathogenesis of arteriosclerosis and cardiovascular disease. *Prev Med, 8:* 379-389, 1979.

Schilling, R.S.F.: *Occupational Health Practice.* London, Butterworths, 1973.

Schleyer, H., Cooper, D.Y., Rosenthal, O., and Cheung, P.: The heme protein P-450. A study of the formation of its ferrous carbon monoxide complex. *Biophys J, 21:* 139, 1978.

Schmeltz, I., Hoffmann, D., and Wynder, E.L.: The influence of tobacco smoke on indoor atmospheres. Part I. An overview. *Prev Med, 4:* 66-82, 1975.

Schneiderman, G., and Goldstick, T.K.: Carbon monoxide-induced arterial wall hypoxia and atherosclerosis. *Atherosclerosis, 30:* 1-15, 1978.

Scholander, P.F., and Roughton, F.J.W.: A simple micro-gasometric method of estimating carbon monoxide in blood. *J Ind Hyg Toxicol, 24:* 218-221, 1942.

Scholander, P. F., and Roughton, F.J.W.: Micro-gasometric estimation of the blood gases. II. Carbon monoxide. *J Biol Chem, 148:* 551-563, 1943.

Schulte, J.H.: Effects of mild carbon monoxide intoxication. *Arch Environ Health, 7:* 524-530, 1963.

Schulte, J.H.: Sealed environment in relation to health and disease. *Arch Environ Health, 8:* 438-452, 1964.

Schwetz, B.A., Leong, B.K., and Staples, R.E.: Teratology studies on inhaled carbon monoxide and imbibed ethanol in laboratory animals. *Teratology, 11:* 33A, 1975.

Schwetz, B.A., Smith, F.A., Leong, B.K., and Staples, R.E.: Teratogenic potential of inhaled carbon monoxide in mice and rabbits. *Teratology, 19:* 385-392, 1979.

Sealander, J.A.: Survival of *Peromyscus* in relation to environmental temperature and acclimation at high and low temperature. *Am Midland Nat, 46:* 257-263, 1951.

Sebben, J., Pimm, P., and Shephard, R.J.: Cigarette smoke in enclosed public facilities. *Arch Environ Health, 32:* 53-58, 1977.

Sedlak, J., Simkova, T., Durasko, L., et al.: Prevention, diagnosis and treatment of carbon monoxide poisoning in an East-Slovakian ironworks. *Resuscitation, 6:* 155-161, 1978.

Segal, H.: Concorde emissions at Dulles International Airport. Results of measurements. *J Air Pollut Control Assoc, 28:* 1039-1042, 1978.

Seiff, H.E.: *CO as an Indicator of Cigarette Caused Pollution Levels in Inter-City Buses..* Washington, D.C., U.S. Govt. Print. Office (Report 1-11-1973), 1973.

Seiler, W.: Anthropogenic and natural sources of CO and H_2. Paper presented at 85th National Meeting of American Institute of Chemical Engineers, Philadelphia, Pennsylvania, 1978.

Seiler, W., and Junge, C.: Carbon monoxide in the atmosphere. *J Geophys Res, 75:* 2217-2226, 1970.

Sendroy, J., and Liu, S.H.: Gasometric determination of oxygen and carbon monoxide in blood. *J Biol Chem, 89:* 133-152, 1930.

Seppänen, A.: Physical work capacity in relation to carbon monoxide inhalation and tobacco smoking. *Ann Clin Res, 9:* 269-274, 1977.

Seppänen, A.: Smoking in closed spaces and its effects on carboxyhemoglobin saturation of smoking and non-smoking subjects. *Ann Clin Res, 9:* 281-283, 1977.

Seppänen, A., Hakkinen, V., and Tenuku, M.: Effect of gradually increasing carboxyhaemoglobin saturation on visual perception and psychomotor performance of smoking and non-smoking subjects. *Ann Clin Res., 9:* 314-319, 1977.

Seppänen, A., and Uusitalo, A.J.: Carboxyhaemoglobin saturation in relation to smoking and various occupational conditions. *Ann Clin Res, 9:* 261-268, 1977.

Shephard, R.J.: Breath-holding measurement of carbon monoxide diffusion capacity. Comparison of a field test with steady-state and other methods of measurement. *J Physiol, 141:* 408-419, 1958.

Shephard, R.J.: Physical performance in Mexico City. In Goddard, R. (Ed.): *International Symposium on the Effects of Altitude on Physical Performance.* Chicago, Athletic Institute, 1967, pp. 132-134.

Shephard, R.J.: Human endurance and the heart at altitude. *J Sports Med Phys Fitness, 10:* 72-83, 1970.

Shephard, R.J.: The influence of small doses of carbon monoxide upon heart rate. *Respiration, 29:* 516-521, 1972.

Shephard, R.J.: Objectives of human air pollution studies. In Shephard, R.J., Pengelly, D., and Silverman, F. (Eds.): *Proceedings of 1st Canadian Conference on Research in Air Pollution.* Gravenhurst, Ontario, York-Toronto Lung Association, Toronto, 1974a, pp. 150-153.

Shephard, R.J.: *Men at work. Applications of Ergonomics to Performance and Design.* Springfield, Illinois, Charles C Thomas, 1974b.

Shephard, R.J.: *Endurance Fitness.* 2nd ed. Toronto, University of Toronto Press, 1977.

Shephard, R.J.: *Human Physiological Work Capacity.* London, Cambridge University Press, 1978.

Shephard, R.J.: *An Ambient Air Quality Objective for Carbon Monoxide.* Ottawa: Health and Welfare, Canada, 1980.

Shephard, R.J.: *The Risks of Passive Smoking.* London, Croom Helm, 1982a.

Shephard, R.J.: *Physiology and Biochemistry of Exercise.* New York, Praeger, 1982b.

Shephard, R.J., Turner, J.E., Carey, G.C.R., and Phair, J.J.: Correlation of pulmonary function and domestic micro-environment. *J Appl Physiol, 15:* 70-76, 1960.

Shepherd, M.: Rapid determination of small amounts of carbon monoxide. *Anal Chem 19:* 77-81, 1947.

Shepherd, M., Schumann, S., and Kilday, M.V.: Determination of carbon monoxide in air pollution studies. *Anal Chem, 27:* 380-383, 1955.

Shields, C.E.: Elevated carbon monoxide levels from smoking in blood donors. *Transfusion, 11:* 89-93, 1971.

Shillito, F.H., Drinker, C.K., and Shaugnessy, T.J.: The problem of nervous and mental sequelae in carbon monoxide poisoning. *JAMA, 106:* 669-674, 1936.

Shimp, D.M., Blumrosen, A.W., and Finifter, S.B.: How to protect your health at work. Salem, New Jersey, Environmental Improvement Associates, 1976.

Shirling, F.C.A.: Breathing apparatus in the British Fire Service. In Davies, C.N. (Ed.): *Design and Use of Respirators.* Oxford, Pergamon Press, 1961, pp. 136-139.

Shul'ga, T.: New data for the hygienic evaluation of carbon monoxide in atmospheric air. In Levine, B.S. (Ed.): *U.S.S.R. Literature on Air Pollution and Related Occupational Diseases. Vol. 9.* Washington, D.C., U.S. Govt. Print. Office, 1961, pp. 73-81.

Shults, W.D., and Beauchamp, J.J.: Statistically based air-quality indices. In Thomas, W.A. (Ed.): *Indicators of Environmental Quality.* New York, Plenum Press, 1972, pp. 199-209.

Shy, C.M.: Human health consequences of nitrogen dioxide. In *Proceedings of the Conference on Health Effects of Air Pollutants.* Washington, D.C.: U.S. Govt. Print. Office (serial 93-15), 1973, pp. 363-405.

Siegel, S.M., Renwick, G., and Rosen, L.A.: Formation of carbon monoxide during seed germination and seedling growth. *Science, 137:* 683-684, 1962.

Sievers, R.F., Edwards, T.I., Murray, A.L., and Schrenk, H.H.: Effect of exposure to known concentrations of carbon monoxide: a study of traffic officers stationed at the Holland Tunnel for 13 years. *JAMA, 118:* 585-588, 1942.

Siggaard-Andersen, J., Petersen, F.B., Hansen, J.I., and Mellemgaard, K.: Plasma volume and vascular permeability during hypoxia and carbon monoxide exposure. *Scand J Clin Lab Invest, 22, Suppl. 103:* 39-48, 1968.

Silk, S.J.: The threshold limit value for carbon monoxide. *Ann Occup Hyg, 18:* 29-35, 1975.

Silverman, L., and Gardner, G.R.: Potassium pallado-sulfite method for carbon monoxide detection. *Am Ind Hyg Assoc J, 26:* 97-105, 1965.

Sirs, J.A.: The effect of temperature on the reaction of carbon monoxide with oxygenated haemoglobin. *J Physiol, 260:* 37-44, 1976.

Sjöstrand, T.: A method for the determination of carboxyhaemoglobin concentrations by analysis of alveolar air. *Acta Physiol Scand, 16:* 201-210, 1948a.

Sjöstrand, T.: A method for the determination of the total hemoglobin content of the body. *Acta Physiol Scand, 16:* 211-231, 1948b.

Sjöstrand, T.: Endogenous formation of carbon monoxide in man under normal and pathological conditions. *Scand J Clin Lab Invest, 1:* 201-214, 1949.

Sjöstrand, T.: Endogenous formation of carbon monoxide: the CO concentration in the inspired and expired air of hospital patients. *Acta Physiol Scand, 22:* 137-141, 1951.

Sjöstrand, T.: The formation of carbon monoxide by the decomposition of haemoglobin in vivo. *Acta Physiol Scand, 26:* 338-344, 1952a.

Sjöstrand, T.: The formation of carbon monoxide by in vitro decomposition of haemoglobin in bile pigments. *Acta Physiol Scand, 26:* 328-333, 1952b.

Sjöstrand, T.: The formation of carbon monoxide in the living organism—a factor to be considered in space flight. Reprinted from *Proceedings of Second International Symposium on Man in Space,* Paris, 1965.

Slusher, G.R.: *Analysis of Engine-Exhaust System Failures in General Aviation Aircraft.* Tech. Rept. FAA ADS 29. Oklahoma City: U.S. Federal Aviation Administration, 1964, pp. 1-18.

Small, K.A., Radford, E.P., Frazier, J.M., et al.: A rapid method for simultaneous measurement of carboxy- and methemoglobin in blood. *J Appl Physiol, 31:* 154-160, 1971.

Smialek, M.: Cytochrome oxidase activity in isolated giant cell of the reticular formation from the rat brain stem following carbon monoxide intoxication (Cartesian Micro-Diver Technique). *Neuropat Pol, 12:* 397-401, 1974.

Smith, E., McMillan, E., and Mack, L.: Factors influencing the lethal action of illuminating gas. *J Ind Hyg, 17:* 18-20, 1935.

Smith, G.: The treatment of carbon monoxide poisoning with oxygen at 2 atmospheres pressure. In Cunningham, D.J.C., and Lloyd, B.B. (Eds.): *The Regulation of Human Respiration.* Oxford, Blackwell, 1963, pp. 569-570.

Smith, J.S., and Brandon, S.: Morbidity from acute carbon monoxide poisoning at three-year follow-up. *Br Med J,* (i): 318-321, 1973.

Smith, N.J.: Evaluation of both outdoor and indoor exposure to carbon monoxide and their health significance in Calgary, Alberta, Canada. Calgary, University of Calgary Environmental Sciences Centre, 1976a.

Smith, N.J.: Occupational and community exposure to carbon monoxide in Calgary, Alberta. Part 2: Motor vehicle orientated occupations. Calgary, University of Calgary Environmental Sciences Centre, 1976b.

Smith, N.J.: End-expired air technic for determining occupational carbon monoxide exposure. *J Occup Med, 19:* 766-769, 1977.

Smith, P.W., Crane, C.R., Sanders, D.C., et al.: Effects of exposure to carbon monoxide and hydrogen cyanide. In *Physiological and Toxicological Aspects of Combustion Products.* NAS/NRC International Symposium, Salt Lake City, 1976, pp. 75-88.

Smith, R.G.: Tentative method for constant pressure volumetric gas analysis for oxygen, carbon dioxide, carbon monoxide, nitrogen, hydrocarbons. *Health Lab Sci, 12:* 177-181, 1975.

Smith, R.N., Swineheart, J., and Lesnini, D.G.: Chromatographic analysis of gas mixtures containing nitrogen, nitrous oxide, nitric oxide, carbon monoxide and carbon dioxide. *Anal Chem, 30:* 1217-1218, 1958.

Smith, W.S., and Gruber, C.W.: *Atmospheric Emissions from Coal Combustion—an Inventory Guide. Environmental Health Series, Air Pollution.* Cincinnati, U.S. Govt. Print. Office (DHEW Publication 999-AP-24), 1966.

Sofuluwe, G.: Smoke pollution in dwellings of infants with bronchopneumonia. *Arch Environ Health, 16:* 670-672, 1968.

Sollberger, A.: Biological measurements in time, with particular reference to synchronization mechanisms. *Ann NY Acad Sci, 138:* 561-599, 1967.

Song, T.M.K., and Eagan, C.J.: Effect of running on rate of clearance of carbon monoxide from the body. *Med Sci Sports, 10:* 60, 1978.

Speert, H.: Noxious vapours in aircraft cabins. *Occup Med, 2:* 101-115, 1946.

Spengler, J.D., Stone, K.R., and Lilley, F.W.: High carbon monoxide levels measured in enclosed skating rinks. *J Air Pollut Control Assoc, 28:* 776-779, 1978.

Sprout, W.L., Neeld, W.E., and Woessner, W.W.: Management of chemical cyanosis to oxygen saturation readings. *Arch Environ Health, 30:* 302-306, 1975.

Srch, M.: Ueber die Bedeutung des Kohlenoxyds beim Zigarettenrauchen im Personenkraftwageninnern. *Dtsch Z Ges Med, 60:* 80-89, 1967.

Steinfeld, J.L.: *The Health Consequences of Smoking. A Report of the Surgeon General.* Washington, D.C., U.S. Govt. Print. Office (DHEW Publication), 1972.

Stembridge, V.A., and Goldbaum, L.R.: *The Role of the Armed Forces Institute of Pathology in Tissue Toxicity Study.* Aviation Medicine Symposium, 1957. Cited by Cooper, A.G. (1966).

Stender, S., Åstrup, P., and Kjeldsen, K.: The effect of carbon monoxide on cholesterol in the aortic wall of rabbits. *Atherosclerosis, 28:* 357-367, 1977.

Sterling, T.D., and Kobayashi, D.M.: Exposure to pollutants in enclosed "living spaces." *Environ Res, 13:* 1-35, 1977.

Stern, A.C.: Basis for criteria and standards. *J Air Pollut Control Assoc, 15:* 281-283, 1965.

Stern, A.C., Wohlers, H.C., Boubel, R.W., and Lowry, W.P.: *Fundamentals of Air Pollution.* New York, Academic Press, 1973.

Stetter, J.R., and Blurton, K.F.: Portable high temperature catalytic reactor: application to air pollution monitoring instrumentation. *Rev Sci Instrum, 47:* 691-694, 1976.

Stevens, C.M., Krout, L., Walling, D., et al.: The isotopic composition of atmospheric carbon monoxide. *Earth and Plant Sci Letters, 16:* 147-165, 1972.

Stevenson, M.F., Cooper, G.L., and Chenoweth, M.B.: Effect on carboxyhemoglobin of exposure to aerosol spray paints with methylene chloride. *Clin Toxicol, 12:* 551-561, 1978.

Stewart, C.P., and Stolman, A.: *Toxicology. Mechanisms and Analytical Methods. Vol II.* New York, Academic Press, 1961.

Stewart, R.D.: The effects of low concentrations of carbon monoxide in man. *Scand J Respir Dis (Suppl), 91:* 56-62, 1974.

Stewart, R.D.: The effect of carbon monoxide on humans. *J Occup Med, 18:* 304-309, 1976.

Stewart, R.D., Peterson, J.E., Bachand, R.T., et al.: Experimental human exposure to carbon monoxide. *Arch Environ Health, 21:* 154-164, 1970.

Stewart, R.D., Fisher, T.N., Hosko, M.J., et al.: Experimental human exposure to methylene chloride. *Arch Environ Health, 25:* 342-348, 1972a.

Stewart, R.D., Fisher, T.N., Hosko, M.J., et al.: Carboxyhemoglobin elevation after exposure to dichloromethane. *Science, 176:* 295-296, 1972b.

Stewart, R.D., Newton, P.E., Hosko, M.J., and Peterson, J.: Effect of carbon monoxide on time perception. *Arch Environ Health, 27:* 155-160, 1973.

Stewart, R.D., Baretta, E.D., Platte, L.R., et al.: *"Normal" Carboxyhemoglobin Levels of Blood Donors in the United States.* Report ENVIRON-MED-MCW CRC-COHb-73-1, Medical College of Wisconsin, 1973.

Stewart, R.D., Peterson, J.E., Fisher, T.N., et al.: Experimental human exposure to high concentrations of carbon monoxide. *Arch Environ Health, 26:* 1-7, 1973.

Stewart, R.D., Baretta, E.D., Platte, L.R., et al.: Carboxyhemoglobin levels in American blood donors. *JAMA, 229:* 1187-1195, 1974.

Stewart, R.D., et al.: *Exposure of Humans to Carbon Monoxide Combined with Ingestion of Ethyl Alcohol and the Comparison of Human Performance when Exposed for Varying Periods of Time to Carbon Monoxide.* Report to Coordinating Research Council and Environmental Protection Agency on Contract CRC-APRAC, Project CAPM-3-68.

Medical College of Wisconsin, 1974, pp. 1-39.

Stewart, R.D., Newton, P.E., Hosko, M.J., et al.: The effect of carbon monoxide on time perception, manual coordination, inspection and arithmetic. In Weiss, B., and Laties, V.G. (Eds.): *Behavioral Toxicology*. New York, Plenum Press, 1975, pp. 29-55.

Stewart, R.D., and Hake, C.L.: Paint remover hazard. *JAMA, 235:* 398-401, 1976.

Stewart, R.D., Hake, C.L., Wu, A., et al.: Carboxyhemoglobin trend in Chicago blood donors 1970-1974. *Arch Environ Health, 31:* 280-285, 1976.

Stewart, R.D., Stewart, R.S., Stamm, W., and Seelen, R.P.: Rapid estimation of carboxyhaemoglobin level in firefighters. *JAMA, 235:* 390-392, 1976.

Strange, J.P., Ball, K.E., and Barnes, D.O.: Continuous parts per billion recorder for air contaminants. *J Air Pollut Control Assoc, 10:* 423-426, 1960.

Stryer, L.: *Biochemistry*, 2nd ed. San Francisco, W.H. Freeman, 1981.

Stupfel, M., and Bouley, G.: Physiological and biochemical effects on rats and mice exposed to small concentrations of carbon monoxide for long periods of time. *Ann NY Acad Sci, 174:* 342-368, 1970.

Surgeon General. *Smoking and Health*. Washington, D.C., U.S. Govt. Print. Office (DHEW Publication PHS-79-50066), 1979.

Surgeon General. *The Health Consequences of Smoking for Women*. Washington, D.C., U.S. Govt. Print. Office, 1980.

Sutton, W.R., Russell, M.A.H., Feyerabend, C., and Saloojee, Y.: Smokers' response to dilution of smoke by ventilated cigarette holders. In Thornton, R.E. (Ed.): *Smoking Behavior. Physiological and Psychological Influences*. Edinburgh, Churchill-Livingstone, 1978, pp. 330-335.

Sweet, A.H., Steigerwald, B.J., and Ludwig, J.H.: The need for a pollution-free vehicle. *J Air Pollut Control Assoc, 18:* 111-113, 1968.

Swinnerton, J.W., Linnenbom, V.J., and Lamontagne, R.A.: The ocean: a natural source of carbon monoxide. *Science, 167:* 984-986, 1970.

Syvertsen, G.R., and Harris, G.A.: Erythropoietin production in dogs exposed to high altitude and carbon monoxide. *Am J Physiol, 225:* 293-299, 1973.

Szadkowski, D., Harke, H.P., and Angerer, J.: Kohlenmonoxidbelastung durch Passivrauchen in Buroraumen. *Innere Medizin, 3:* 310-313, 1976.

Szulcewski, D.H., and Higuchi, T.: Gas chromatographic separation of some permanent gases on silica gel at reduced temperatures. *Anal Chem, 29:* 1541-1543, 1957.

Tanaka, M.: Studies on the etiological mechanism of fetal developmental disorders caused by maternal smoking during pregnancy. *J Jap Obstet Gynecol, 17:* 1107-1114 (English translation in *J Obstet Gynecol Soc, 14:* 45), 1967.

Teague, M.C.: The determination of carbon monoxide in air contaminated with motor exhaust gas. *J Ind Eng Chem, 12:* 964-968, 1920.

Teige, B., Lundevall, J., and Fleischer, E.: Carboxyhaemoglobin concentrations in fire victims and in cases of fatal carbon monoxide poisoning. *Z Rechtsmed, 80:* 17-21, 1977.

Tenney, S.M., and Ou, L.C.: Physiological evidence for increased tissue capillarity in rats acclimatized to high altitude. *Respir Physiol, 8:* 137-150, 1970.

Terrill, J.B., Montgomery, R.R., and Reinhardt, C.F.: Toxic gases from fires. *Science, 200:* 1343-1347, 1978.

Theodore, J.: Angina pectoris and carbon monoxide. (Letter). *Ann Int Med, 78:* 455, 1973.

Theodore, J., O'Donnell, R.D., and Back, K.C.: Toxicological evaluation of carbon monoxide in humans and other mammalian species. *J Occup Med, 13:* 242-255, 1971.

Thomas, H.M., Lefrak, S.S., Irwin, R.S., et al.: The oxyhemoglobin dissociation curve in health and disease: role of 2,3 diphosphoglycerate. *Am J Med, 57:* 331-348, 1974.

Thomas, M.F., and Penney, D.G.: Hematologic responses to carbon monoxide and altitude:

a comparative study. *J Appl Physiol, 43:* 365-369, 1977.

Thomsen, H.K.: Carbon monoxide induced atherosclerosis in primates. *Atherosclerosis, 20:* 233-240, 1974.

Thomsen, H.K., and Kjeldsen, K.: Threshold limit for carbon monoxide-induced myocardial damage. An electron microscopic study in rabbits. *Arch Environ Health, 29:* 73-78, 1974.

Tiao, G.C., Box, G.E.P., and Hamming, W.J.: A statistical analysis of the Los Angeles ambient carbon monoxide data 1955-1972. *J Air Pollut Control Assoc, 25:* 1129-1136, 1975.

Tillmanns, H., Sarma, I.S.M., Seeler, K., et al.: Lipid metabolism in perfused human coronary arteries and saphenous veins. In Schettler, G., and Weizel, A. (Eds.): *Atherosclerosis Vol. 3.* New York, Springer-Verlag, 1975, pp. 118-125.

Tobias, C.A., Lawrence, J.H., Roughton, F.J.W., et al.: The elimination of carbon monoxide from the human body with reference to the possible conversion of CO to CO_2. *Am J Physiol, 145:* 253-263, 1945.

Toliver, W.H., and Morris, M.L.: Chemical analysis of permanent and organic gases in a 30-day manned experiment. *Aerospace Med, 37:* 233-238, 1966.

Topping, D.L.: Metabolic effects of carbon monoxide in relation to atherogenesis. *Atherosclerosis, 26:* 129-137, 1977.

Torbati, I.D., and Har-Kedar, I.: Carboxyhemoglobin levels in blood donors in relation to cigarette smoking and to occupational exposure to carbon monoxide. *Isr J Med Sci, 10:* 241-244, 1974.

Traystman, R.J., and Fitzgerald, R.S.: Cerebral circulatory responses to hypoxic hypoxia and carbon monoxide hypoxia in carotid baroreceptor and chemoreceptor denervated dogs. *Acta Neurol Scand 56, Suppl 64:* 294-295, 1977.

Trompeo, G., Turletti, G., and Giarusso, O.T.: Concentration of carbon monoxide in underground garages. *Rass Med Ind, Turin, 33:* 392-393, 1964.

Tumasonis, C.F., and Baker, F.D.: Influence of carbon monoxide upon some respiratory enzymes of the chick embryo. *Bull Environ Contam Toxicol, 8:* 113-119, 1972.

Turner, D.M.: Carbon monoxide, tobacco smoking and the pathogenesis of atherosclerosis. *Prev Med, 8:* 303-309, 1979.

Twiss, S.B., Teague, D.M., Bozek, J.W., and Sink, M.V.: Application of infra-red spectroscopy to exhaust gas analysis. *J Air Pollut Control Assoc, 5:* 75-83, 1955.

Ury, H.K., Perkins, N.M., and Goldsmith, J.R.: Motor vehicle accidents and vehicular pollution in Los Angeles. *Arch Environ Health, 25:* 314-322, 1972.

United States Bureau of Mines: *Report of U.S. Bureau of Mines.* New York, J Am Soc Heat Vent Engrs, 1927.

Valentin, H., Bost, H.P., and Wawra, E.: Das Passivrauchen am Arbeitsplatz—eine Gesundheitsschadigung? *Zentralbl Bakteriol, 167:* 405-434, 1978.

Vanderslice, J.: State laws on smoking in public places. *Am Lung Assoc Bull, 62:* 8-10, 1976.

Van Slyke, D.D., and Neill, J.M.: Determination of bases in blood and other solutions by vacuum extraction and manometric measurement. *J Biol Chem, 61:* 523-584, 1924.

Van Slyke, D.D., Hiller, A., Weisiger, J.R., and Cruz, W.O.: Determination of carbon monoxide in blood and of total and active hemoglobin by carbon monoxide capacity. Inactive hemoglobin and methemoglobin contents of normal human blood. *J Biol Chem, 166:* 121-148, 1946.

Vaughan, B.E., and Pace, N.: Changes in myoglobin content of the high altitude acclimatized rat. *Am J Physiol, 185:* 549-556, 1956.

Vernon, H.M.: Prevention of accidents. *Br J Ind Med, 2:* 1-9, 1945.

Vernot, E.H., MacKenzie, W.F., MacEwen, J.D., et al.: *Hematological Effects of Long-Term Continuous Animal Exposure to Carbon Monoxide.* U.S. Aerospace Medical Research Laboratory Report AMRL-TR-70-102, Paper #1, 1970.

Virtamo, M., and Tossaveinen, A.: Carbon monoxide in foundry air. *Scand J Work Environ Health, Suppl. 1:* 37-41, 1976.

Vogel, J.A., and Gleser, M.A.: Effect of carbon monoxide on oxygen transport during exercise. *J Appl Physiol, 32:* 234-239, 1972.

Vogel, J.A., Gleser, M.A., Wheeler, R.C., and Whitten, B.K.: Carbon monoxide and physical work capacity. *Arch Environ Health, 24:* 198-203, 1972.

Vogel, J.A., Hartley, L.H., and Cruz, J.C.: Cardiac output during exercise in altitude natives at sea level and high altitude. *J Appl Physiol, 36:* 173-176, 1974.

Vogt, T.M., Selvin, S., Widdowson, G., and Hulley, S.B.: Expired air carbon monoxide and serum thiocyanate as objective measures of cigarette exposure. *Am J Public Health, 67:* 545-549, 1977.

Vollmer, E.P., King, B.G., Birren, J.E., and Fischer, M.B.: The effects of CO on three types of performance; at simulated altitudes of 10,000 and 15,000 feet. *J Exp Psychol, 36:* 244-251, 1946.

Von Lüpke, H., and Schmidt, K.: Combined carbon monoxide/cyanide intoxication due to combustion of plastic material. *Z Rechtsmed, 79:* 69-71, 1977.

Wagner, J.A., Horvath, S.M., and Dahms, T.E.: Carbon monoxide elimination. *Respir Physiol, 23:* 41-47, 1975.

Wagner, J.A., Horvath, S.M., and Dahms, T.E.: Cardiovascular adjustments to carbon monoxide exposure during rest and exercise in dogs. *Environ Res, 15:* 368-374, 1978.

Wahl, F.: Uber den Gehalt des Tabakrauches an Kohlenoxyd. *Pfluegers Arch, 88:* 262-285, 1899.

Wakeham, H.R.: Environmental carbon monoxide from cigarette smoke—a critique. *Prev Med, 6:* 526-534, 1977a.

Wakeham, H.R.: Environmental carbon monoxide from cigarette smoking. In *Proceedings of 6th International Tobacco Scientific Congress.* Tokyo, Japan Tobacco and Public Salt Corporation, 1977b, pp. 93-101.

Wald, G., and Allen, D.W.: Equilibrium between cytochrome oxidase and carbon monoxide. *J Gen Physiol, 40:* 593-609, 1967.

Wald, N., Howard, S., Smith, P.G., and Kjeldsen, K.: Association between atherosclerotic diseases and carboxyhaemoglobin levels in tobacco smokers. *Br Med J,* (i): 761-765, 1973.

Wald, N., and Smith, P.G.: Carbon monoxide and smoking. *Lancet,* (i): 750, 1974.

Wald, N., Idle, M., and Smith, P.G.: Carboxyhaemoglobin levels in smokers of filter and plain cigarettes. *Lancet,* (i): 111-112, 1977.

Wald, N.J.: Carbon monoxide as an aetological agent in arterial disease—some human evidence. In Wynder, E.L., Hoffmann, D., and Gori, G.B. (Eds.): *Modifying the Risk for the Smoker. Vol. 1. Proceedings of the 3rd World Conference on Smoking and Health.* Washington, D.C., U.S. Govt. Print. Office (DHEW Publication NIH 76-1221), 1975, pp. 349-361.

Walford, J.: Field surveys and sampling methods. In Schilling, R.S.F. (Ed.): *Occupational Health Practice.* London, Butterworths, 1973, pp. 190-210.

Waller, J.A., and Thomas, K.: Carbon monoxide, smoking and fatal highway crashes. *Proceedings 15th Conference, American Association for Automotive Medicine.* Colorado Springs, 1971.

Walsh, M., and Nussbau, B.: Who's responsible for emissions after 50,000 miles? *Automotive Eng, 86:* 32-35, 1978.

Wanstrup, J., Kjeldsen, K., and Åstrup, P.: Acceleration of spontaneous changes in rabbit aorta by a prolonged, moderate carbon monoxide exposure. *Acta Pathol Microbiol Scand, 75:* 353-362, 1969.

Warburg, O.: Uber die Wirkung des Kohlenoxyds auf den Stoffwechsel der Hefe. *Biochem Z,*

177: 471-486, 1926.

Warwick, K.M., and Eysenck, H.J.: The effects of smoking on the CFF threshold. *Life Sci, 2:* 219-223, 1963.

Weber, A., Jeremini, C., and Grandjean, E.: Effects of low carbon monoxide concentrations on flicker fusion frequency and on subjective feelings. *Int Arch Occup Environ Health, 36:* 87-103, 1975.

Weinstock, B.: Carbon monoxide: Residence time in the atmosphere. *Science, 166:* 224-225, 1969.

Weinstock, B., and Niki, H.: Carbon monoxide balance in nature. *Science, 176:* 290-292, 1972.

Weinstock, B., and Chang, T.Y.: The global balance of carbon monoxide. *Tellus, 26:* 108-115, 1974.

Weir, F.W., Mehta, M.M., Johnson, D.F., et al.: *An Investigation of the Effects of Carbon Monoxide on Humans in the Driving Task.* Ohio State University Research Foundation, January, 1973.

Weiser, P.C., Morrill, C.G., Dickey, D.W., et al.: Effect of low-level carbon monoxide on the adaptation of healthy young men to aerobic work at an altitude of 1,610 m. In Folinsbee, L.J., Wagner, J.A., Borgia, J.F., Drinkwater, B.L., Gliner, J.A., and Bedi, J.F. (Eds.): *Environmental Stress: Individual Human Adaptations.* New York, Academic Press, 1979.

Weiss, H.R., and Cohen, J.A.: Effects of low levels of carbon monoxide on rat brain and muscle tissue PO_2. *Environ Physiol Biochem, 4:* 31-39, 1974.

Wells, L.L.: The prenatal effect of carbon monoxide on albino rats and the resulting neuropathology. *Biologist, 15:* 80-81, 1933.

Wennesland, R.: A new method for the determination of carbon monoxide in blood. *Acta Physiol Scand, 1:* 49-70, 1940.

Werner, A.: A sensitive, rapid infrared method for analysing carbon monoxide in blood. *Scand J Clin Lab Invest, 36:* 203-205, 1976.

Whalen, W.J.: Intracellular PO_2 in heart and skeletal muscle. *Physiologist, 14:* 69-82, 1971.

Wheeler, G. (Chairman of committee appointed by Mining Association of Great Britain): *Record of British Coal-Dust Experiments.* London, Colliery Guardian Company Ltd., 1910.

White, D.C.: Mitochondrial cytotoxicity of carbon monoxide. In *Proceedings of 1975 Symposium Nicotine and Carbon Monoxide.* Lexington, Kentucky, University of Kentucky Press, 1977.

White, J.J.: Carbon monoxide and its relation to aircraft. *U.S. Naval Med Bull. 30:* 515-565, 1932.

Whitehead, T.P., and Worthington, S.: The determination of carboxyhaemoglobin. *Clin Chim Acta, 6:* 356-359, 1961.

World Health Organization: *Atmospheric Pollutants.* WHO Tech. Report *271,* 1964.

World Health Organization: *Health Aspects Related to Indoor Air Quality. Report on a WHO Working Group.* Copenhagen, WHO Regional Office for Europe, 1979a.

World Health Organization: *Environmental Health Criteria 13. Carbon Monoxide.* Geneva, 1979b.

Wilkins, E.T.: Some measurements of carbon monoxide in the air of London. *J R Soc Health, 76:* 677-684, 1956.

Wilks, S.S.: Carbon monoxide in green plants. *Science, 129:* 964-966, 1959.

Wilks, S.S.: Toxic photo-oxidation products in closed environments. *Aerospace Med, 34:* 838-841, 1963.

Wilks, S.S., Tomashefski, J.F., and Clark, R.T.: Physiological effects of chronic exposure to carbon monoxide. *J Appl Physiol, 14:* 305-310, 1959.

Williams, I.R., and Smith, E.: Blood picture reproduction and general condition during daily exposure to illuminating gas. *Am J Physiol, 110:* 611-615, 1935.

Williams, R.: De Sanitate Urbanorum. *Br Med J, (ii) (6205):* 1617-1620, 1979.

Wilson, A.J., and Schaefer, K.E.: Effect of prolonged exposure to elevated carbon monoxide and carbon dioxide levels on red blood cell parameters during submarine patrols. *Undersea Biomed Res, 6, (Submarine Suppl):* 549-556, 1979.

Wilson, D.F., Swinnerton, J.W., and Lamontagne, R.A.: Production of carbon monoxide and gaseous hydrocarbons in seawater: Relation to dissolved organic carbon. *Science, 168:* 1577-1579, 1970.

Wilson, E.D., Gates, I., Owen, H.R., and Dawson, W.T.: Street risk of carbon monoxide poisoning. *JAMA, 87:* 319-320, 1926.

Winneke, G.: Behavioral effects of methylene chloride and carbon monoxide as assessed by sensory and psychomotor performance. In Xintaras, C., Johnson, B., and deGroot, I. (Eds.): *Behavioral Toxicology: Early Detection of Occupational Hazards.* Washington, D.C., U.S. Govt. Printing Office (DHEW Publication NIOSH Report 74-126), 1974, pp. 130-144.

Winneke, G., Fodor, G.G., and Schlipkoter, H.W.: *Effects of Low Level Carbon Monoxide Exposure and Monotony on Vigilance and Vertex-Evoked Potentials.* Paper presented at 4th International Congress on Event Related Slow Potentials of the Brain, Hendersonville, North Carolina, April 4-10, 1976. Cited by Environmental Criteria and Assessment Office (1979).

Wissler, R.W., Vesselinovitch, D., and Getz, G.S.: Abnormalities of the arterial wall and its metabolism in atherogenesis. *Prog Cardiovasc Dis, 18:* 341-369, 1976.

Wohlrab, H., and Ogunmola, G.B.: Carbon monoxide binding studies of cytochrome a, hemes in intact liver mitochondria. *Biochemistry, 10:* 1103-1106, 1971.

Wood, E.N.: Increased incidence of stillbirth in piglets associated with levels of atmospheric carbon monoxide. *Vet Rec, 104:* 283-284, 1979.

Wright, G.R.: Adaptive control in vigilance. Master's thesis, Department of Industrial Engineering, University of Toronto, 1971.

Wright, G.R.: Effects of carbon monoxide on human performance. Doctoral dissertation, University of Toronto, 1976.

Wright, G.R., Randell, P., and Shephard, R.J.: Carbon monoxide and driving skills. *Arch Environ Health, 27:*349-354, 1973.

Wright, G.R., Polk, L., and Shephard, R.J.: *Carbon Monoxide in Dwellings.* Washington, D.C., U.S. Federal Committee, Bureau of Community Environment Management (Report on domestic standards of CO exposure), 1974.

Wright, G.R., Jewczyk, S., Onrot, J., et al.: *Carbon Monoxide in Downtown Toronto.* Willowdale, Ontario, York Toronto Tuberculosis and Respiratory Disease Association, 1974.

Wright, G.R., Baum, J., Cogan, F., and Shephard, R.J.: Smoking and carboxyhemoglobin levels. *Lancet, (i):* 635, 1974.

Wright, G.R., Jewczyk, S., Onrot, J., et al.: Carbon monoxide in the urban atmosphere. Hazards to the pedestrian and the street-worker. *Arch Environ Health, 30:* 123-129, 1975.

Wright, G.R., and Shephard, R.J.: Carbon monoxide, nicotine and the "safer" cigarette. *Respiration, 35:* 40-52, 1978a.

Wright, G.R., and Shephard, R.J.: A note on blood carboxyhemoglobin levels in police officers. *Can J Public Health, 69:* 393-398, 1978b.

Wright, G.R., and Shephard, R.J.: Brake reaction time—effects of age, sex and carbon monoxide. *Arch Environ Health, 33:* 141-150, 1978c.

Wright, G.R., and Shephard, R.J.: Carbon monoxide exposure and auditory duration discrimination. *Arch Environ Health, 33 (5):* 226-235, 1978d.

Wright, W.C. (Trans.): B. Ramazzini's *De Morbis Artificum.* New York, Hafner, 1964.

Wyburn, C.M., Pickford, R.W., and Hirst, R.J.: *Human Senses and Perception.* Toronto, University of Toronto Press, 1964.

Wynder, E.L., and Hoffmann, D.: *Tobacco and Tobacco Smoke. Studies in Experimental Carcinogenesis.* New York, Academic Press, 1967.

Xintaras, C., Johnson, B.L., Ulrich, C.E., et al.: Application of the evoked response technique in air pollution toxicology. *Toxicol Appl Pharmacol, 8:* 77-87, 1966.

Xintaras, C., Johnson, B., and deGroot, I. (Eds.): *Behavioral Toxicology, Early Detection of Occupational Hazards.* Washington, D.C., U.S. Govt. Print. Office (DHEW Publication NIOSH Report 74-126), 1974.

Yabroff, I., Myers, E., Fend, V., et al.: *The Role of Atmospheric Carbon Monoxide in Vehicle Accidents.* Stanford Research Institute, (Report on Contract CRC-APRAC Project CAPM-12-69), 1974, pp. 1-92.

Yamate, N., and Inoue, A.: Continuous analyzer for carbon monoxide in ambient air by electrochemical technique. *J Environ Contr (Japan), 9:* 292-296, 1973.

Yates, M.W.: A preliminary study of carbon monoxide in the home. *J Environ Health, 29:* 413-420, 1967.

Yocom, J.E., Cote, W.A., and Clink, W.L.: Indoor/outdoor air quality relationships. *J Air Pollut Control Assoc, 21:* 251-259, 1971.

Young, I.M., and Pugh, L.G.C.: The carbon monoxide content of foetal and maternal blood. *J Obstet Gynaecol Br Commonw, 70:* 681-684, 1963.

Young, S.H., and Stone, H.L.: Effect of a reduction in arterial oxygen content (carbon monoxide) on coronary flow. *Aviat Space Environ Med, 47:* 142-146, 1976.

Younoszai, M.K., Kacic, A., and Haworth, J.C.: Cigarette smoking during pregnancy: the effect upon the haematocrit and the acid-base balance of the newborn infant. *Can Med Assoc J, 99:* 197-200, 1968.

Younoszai, M.K., and Haworth, J.C.: Placental dimensions and relations in pre-term, term, and growth-retarded infants. *Am J Obstet Gynecol, 103:* 265-271, 1969.

Zebro, T., Littleton, R.J., and Wright, E.A.: Adaptation of mice to carbon monoxide and the effect of splenectomy. *Virchows Arch A, 371:* 35-51, 1976.

Zenz, C.: The epidemiology of carbon monoxide in cardiovascular disease in industrial environments: a review. *Prev Med, 8:* 279-288, 1979.

Zielhuis, R.L.: Tentative emergency exposure limits for sulphur dioxide, sulphuric acid, chlorine and phosgene. *Ann Occup Hyg, 13:* 171-176, 1970.

Zikria, B.A., Weston, T.R., Chodoff, M., and Ferrer, J.M.: Smoke and carbon monoxide poisoning in fire victims. *J Trauma, 12:* 641-645, 1972.

Zorn, H.: The partial oxygen pressure in the brain and liver at subtoxic concentrations of carbon monoxide. *Staub-Reinhalt Luft, 32:* 24-29, 1972.

Zorn, O., and Kruger, P.D.: The problem of chronic carbon monoxide poisoning. *Ind Med Surg, 29:* 580-581, 1960.

INDEX